T0024109

Systems Biology: A Very Short Introduction

VERY SHORT INTRODUCTIONS are for anyone wanting a stimulating and accessible way into a new subject. They are written by experts, and have been translated into more than 45 different languages.

The series began in 1995, and now covers a wide variety of topics in every discipline. The VSI library currently contains over 650 volumes—a Very Short Introduction to everything from Psychology and Philosophy of Science to American History and Relativity—and continues to grow in every subject area.

Very Short Introductions available now:

ABOLITIONISM Richard S. Newman
THE ABRAHAMIC RELIGIONS
   Charles L. Cohen
ACCOUNTING Christopher Nobes
ADAM SMITH Christopher J. Berry
ADOLESCENCE Peter K. Smith
ADVERTISING Winston Fletcher
AERIAL WARFARE Frank Ledwidge
AESTHETICS Bence Nanay
AFRICAN AMERICAN RELIGION
   Eddie S. Glaude Jr
AFRICAN HISTORY John Parker
   and Richard Rathbone
AFRICAN POLITICS Ian Taylor
AFRICAN RELIGIONS
   Jacob K. Olupona
AGEING Nancy A. Pachana
AGNOSTICISM Robin Le Poidevin
AGRICULTURE Paul Brassley and
   Richard Soffe
ALBERT CAMUS Oliver Gloag
ALEXANDER THE GREAT
   Hugh Bowden
ALGEBRA Peter M. Higgins
AMERICAN CULTURAL HISTORY
   Eric Avila
AMERICAN FOREIGN RELATIONS
   Andrew Preston
AMERICAN HISTORY Paul S. Boyer
AMERICAN IMMIGRATION
   David A. Gerber
AMERICAN LEGAL HISTORY
   G. Edward White
AMERICAN NAVAL HISTORY
   Craig L. Symonds

AMERICAN POLITICAL HISTORY
   Donald Critchlow
AMERICAN POLITICAL PARTIES
   AND ELECTIONS L. Sandy Maisel
AMERICAN POLITICS
   Richard M. Valelly
THE AMERICAN PRESIDENCY
   Charles O. Jones
THE AMERICAN REVOLUTION
   Robert J. Allison
AMERICAN SLAVERY
   Heather Andrea Williams
THE AMERICAN WEST Stephen Aron
AMERICAN WOMEN'S HISTORY
   Susan Ware
ANAESTHESIA Aidan O'Donnell
ANALYTIC PHILOSOPHY
   Michael Beaney
ANARCHISM Colin Ward
ANCIENT ASSYRIA Karen Radner
ANCIENT EGYPT Ian Shaw
ANCIENT EGYPTIAN ART AND
   ARCHITECTURE Christina Riggs
ANCIENT GREECE Paul Cartledge
THE ANCIENT NEAR EAST
   Amanda H. Podany
ANCIENT PHILOSOPHY Julia Annas
ANCIENT WARFARE Harry Sidebottom
ANGELS David Albert Jones
ANGLICANISM Mark Chapman
THE ANGLO-SAXON AGE John Blair
ANIMAL BEHAVIOUR
   Tristram D. Wyatt
THE ANIMAL KINGDOM
   Peter Holland

Available soon:

For more information visit our website

www.oup.com/vsi/

Eberhard O. Voit

SYSTEMS BIOLOGY

A Very Short Introduction

Great Clarendon Street, Oxford, OX2 6DP,
United Kingdom

Oxford University Press is a department of the University of Oxford.
It furthers the University's objective of excellence in research, scholarship,
and education by publishing worldwide. Oxford is a registered trade mark of
Oxford University Press in the UK and in certain other countries

First edition published in 2020

Impression: 2

Published in the United States of America by Oxford University Press
198 Madison Avenue, New York, NY 10016, United States of America

British Library Cataloguing in Publication Data

Data available

Library of Congress Control Number: 2019957933

ISBN 978-0-19-882837-2

Printed in Great Britain by
Ashford Colour Press Ltd, Gosport, Hampshire

Contents

Acknowledgments

This little book on systems biology would not have become reality without the wisdom and vision of Oxford's Senior Commissioning Editor Latha Menon, who recognized the rising importance of systems biology. I was honored to receive her invitation to shape the impression of readers interested in this new field and am very grateful to her. My deep appreciation goes to several individuals who read a first draft and provided ample constructive feedback. I would like to thank, in particular, Alik Emelianov, a student at Westminster School in Atlanta, who ensured that the text can be understood by smart high schoolers. Similarly, I greatly appreciate Noune Sarvazyan's and Benedict Voit's numerous suggestions for improved readability of the first draft. My lab colleagues Jacob Davis, Luis Fonseca, and Carla Kumbale kept me on the straight and narrow with a number of loose scientific details. Last not least, Ann Voit very kindly read the whole volume carefully—three times!—in a truly dedicated effort to eradicate unusual formulations. A big thanks to all of them!

List of illustrations

Chapter 1
What is systems biology all about?

Biology is the study of life. Its goal is to gain comprehensive knowledge of how cells and organisms work, individually or in groups. Biology is interested in all aspects of life, from plants and animals to the rich ecosystems of rainforests and oceans, from genes and proteins to cells and organs, from bacteria and viruses to humans. Biological research relies on its own observations, but it has also learned much from chemistry and physics, whose applications to biology eventually became subspecialties in their own right, as the fields of biochemistry and biophysics.

Systems biology is a new specialty area that actually has exactly the same goals and purposes as general biology, namely, to understand how life works. But in contrast to traditional biology, systems biology pursues these goals with a whole new arsenal of tools that come from mathematics, statistics, computing, and engineering, in addition to biology, biochemistry, and biophysics. Systems biology utilizes these tools to determine the specific roles of the many different components that we find in living organisms, how these components interact with each other, and how they all collaborate to create and support life. These components are often molecules, but they may also be smaller, like atoms, electrons, and protons, or much larger, like mitochondria or the nucleus within a cell, cells themselves, organs, or complex mixtures like blood.

A prominent feature of systems biology is its heavy use of the enormous capacity of modern computers to store, manage, analyze, and interpret huge amounts of data. To appreciate why it is necessary to deal with so many data at the same time, let's look at the seemingly simple example of how we get nourishment from food. The digestive system is of course responsible for this process. It consists of distinct parts, all with their own, clearly defined purposes. The teeth and saliva allow us to chew off and swallow food, the stomach churns the food and secretes acid and enzymes that break down the food into its chemical components, and the small and large intestines take up the nutrients before the rest is excreted. This chain of events does not seem to be overly complicated. However, looking more closely, we quickly find that numerous other systems are required for these processes to work properly. Muscles need to move the food and nutrients throughout the digestive system. The cardiovascular system transports the nutrients from the intestines through the bloodstream to even the most remote areas of the body. Seeing or smelling food can make us hungry, so our eyes and nose are involved. The brain and the nervous system tell us when we are hungry and when we have eaten enough and, without our conscious participation, manage the well-coordinated activities of the oesophagus, stomach, and intestines. Moreover, there are the liver, pancreas, and kidneys that all deal with nutrients and the removal of unwanted materials. All of a sudden, digesting food involves a lot of systems that must work together and are tightly and perfectly coordinated to make the process feasible.

The situation is actually even more complicated, because every single one of these systems consists of a variety of subsystems. In particular, we have known for a long time that all organs and tissues contain cells, and these cells have turned out to be very complicated, finely tuned systems in themselves. They contain subsystems and these in turn contain subsystems, and if we step back for a moment, we realize that life is driven by hierarchies of interacting systems, like a whole society of different Russian

Matryoshka nesting dolls. Systems biology tries to understand how each of these systems works and how all of them, within and beyond their own realms, work together.

Just imagine how many different molecules and processes are involved in something like digestion, and you will see very quickly that we need computers that keep thousands of details 'in mind' and don't forget them. The power of computers is only one of the reasons that systems biology has all of a sudden appeared in the limelight, seemingly coming out of nowhere. A second important reason is a large repertoire of novel and different machines and technologies that permit biological experiments and laboratory analyses that were unthinkable even a few decades ago. As just one example, we can determine with a single measurement how much or how little the activity of any or all of our genes is changed in response to some stimulus, such as being cold or hungry.

More important than these technical and computational aspects is the fact that systems biology looks at the living world in a new way that is radically different from traditional biology. The main goal of biology throughout the past century has been the identification and characterization of as many building blocks of life as possible, and we have amassed enormous amounts of knowledge about genes, proteins, metabolites, and other fundamental components. For instance, just in the field of immunology, a new paper is published about every ten minutes; twenty-four hours a day, seven days a week. That is a lot of new information.

Although we have learned much about the molecular building blocks of life, there are still very many things we simply do not understand. Pressing examples include autoimmune and neurodegenerative diseases like rheumatoid arthritis and Alzheimer's disease, whose root causes and details of disease progression remain unclear, in spite of intensive research over many decades.

Systems biology readily recognizes that detailed knowledge of the molecules and structures of life is crucial, and that traditional research absolutely needs to continue, but it stresses the point that this knowledge alone is insufficient. In addition to looking for all individual parts of cells and organisms, we need methods of putting the parts back together. We need new ways of thinking and new methods of analysis that allow us to understand how the building blocks interact and what controls and regulates these interactions.

You may ask: Have we not always put data and information together? The answer is: *Yes* and *No*. *Yes*, we have always asked what different pieces of information collectively mean and whether we might find something new that we had not seen before. But we never had the tools to study huge numbers of observations simultaneously. As a consequence, we often did not even ask questions that modern systems biology attempts to address, especially if they involve many different pieces of information. To appreciate the slow but important changes in the ways biological research has been performed, let us briefly review how we got where we are today in biology and why systems biology adds a genuinely new perspective. How has the study of life changed over time, what do we know and understand, what exactly is missing, and what is yet to be discovered? A whirlwind tour through the evolution of biology will illustrate the roots and contexts from which the field of systems biology has emerged.

Biology is probably almost as old as humankind. Early forays into manipulating the living world around us included the domestication of wolves, which scientists think occurred between 20,000 and 30,000 years ago, as well as the beginnings of farming, maybe 12,000 years ago. As early as about 7,000 years ago, the first beer was brewed. Agriculture did not just mean planting and harvesting, it also required a basic understanding of seasons and life cycles, along with rudimentary record keeping of how much and how quickly the tribe grew or shrank, in order to estimate how much food needed to be secured to survive the winter or other adverse conditions.

A very important driver of biological knowledge has always been medicine. Chinese and Egyptian medicine took root about 5,000 years ago, presumably based on animal observations and on trial and error with people. Much later, we know of Hippocrates (*c.*460–*c.*370 BC), the Greek 'father of medicine,' whom one could also consider an early systems biologist, because he treated the human body as one fully connected system. He taught that health was determined by the whole of an organism and its surroundings, including diet, lifestyle, family history, and environmental factors. Not having knowledge of chemistry, he proposed that health was the result of the right balance of four bodily fluids, called *humors*, and identified these as blood, black bile, yellow bile, and phlegm. In some sense, we have come full circle today, emphasizing the importance of lifestyle and a balanced diet for healthy living, while considering environmental exposures and a 'bad' family history as risk factors. However, in comparison with the present, Hippocrates' 'system biology' had no foundation in physics, chemistry, and biology, whereas we now have extensive information about the processes occurring inside the body, and the field of epigenetics is beginning to reveal the interactions between genes, environment, and lifestyle on a molecular level.

As the father of Western thinking, Aristotle (384–322 BC) created the foundation of the philosophy of science and had a special interest in biology. His approach was quite modern sounding: he proposed to investigate facts as opposed to religious beliefs, to determine whether structures observed in one organism, such as a bird, were also found in other birds and maybe even in fish, and what the reason for their existence might be. He was also the first documented scientist to propose what has become a popular cliché and is also a central theme in modern systems biology, namely, that the whole can be greater than the sum of its parts.

This notion of *synergism* ('working together') is a critical observation that transcends all of biology and will come up many times in this book.

After the demise of the Greek and Roman Empires, the medieval Islamic world engaged in astronomy, mathematics, and medicine, while the Western world entered the dark ages, where there were no major breakthroughs in the sciences, from what we know. Instead, mysticism and alchemy were on the rise, with their quest for a universal elixir for eternal youth and the goal of making gold out of mercury or lead. The Scientific Revolution of the 17th century brought real change with two great advances in biology. First, the microscope was invented and paved the way to an entirely new world of life: microbiology. For us today, it is hard to imagine a world without recognizing microbes, but nobody at the time had as much as an inkling of the existence of the incredible variety and complexity of this 'invisible world.' The second important novelty in the 17th century was a firm rule set for valid research, called the *scientific method*. This method demands that exact science follow well-defined steps of formulating a hypothesis, testing this hypothesis with experiments, analyzing the results, and, based on the results and their interpretation, formulating a new hypothesis. The scientific method is still a cornerstone of research today, although there are some recent alternative strategies that we will discuss in Chapters 3 and 4. During the 18th century, enlightenment, rational thought, and science had finally displaced alchemy. Biology in the 19th century was dominated by physiology, which focused on the various bodily systems governing health and disease, such as the nervous system, the cardiovascular system, the digestive system, and so on. Clearly, the physiology of the 19th century was a direct precursor of systems biology.

Moving into the 20th century, and pursuing the concepts of physiology, an old notion of the philosopher René Descartes re-emerged, which asserted that living organisms were merely complicated machines. Specifically, this assertion suggested that to understand an organism, one had to understand its organs; to understand these, one had to understand tissues and cells; and understanding cells required knowledge of all molecular components. This strategy of increasingly more detailed investigation is called *reductionism*, because it attempts to reduce life all the way down to its ultimate building blocks. The Holy Grail of reductionism is the identification and characterization of all parts of a cell, collectively called its *molecular inventory*. The quest for identifying this inventory ushered in the field of molecular biology, which became the undisputed biological highlight of the 20th century and constitutes the second important root of today's systems biology, complementing physiology. Scientists learned about DNA and its fundamental role in genetics, and invented ever more sophisticated methods to characterize genes, proteins, and metabolites (natural chemical compounds). In parallel, methods like electron microscopy and molecular imaging began to make structures inside cells visible with a resolution unbelievable even a few decades ago. There is absolutely no doubt that the strategy of reductionism has been extremely successful.

A major breakthrough in molecular biology, with manifestations in many areas, happened around the year 2000: it became possible to parallelize and automate many measurement processes that had formerly required a lot of effort. At the time, the scientific community was already in the process of sequencing essentially all genes in a number of organisms, including humans. And since the genes are the carriers of our own blueprint and contain much of the information that makes us who we are, a big piece of the molecular inventory puzzle had been solved. With the new techniques, and heavily supported by robots, it became possible to measure to what degree specific genes were actually expressed in a sample of cells. Other impressive *high-throughput* methods, like mass spectrometry, flow and mass cytometry, as well as molecular imaging, began to allow the characterization of very large inventories of proteins and metabolites, eventually even in single cells. Chapter 3 presents more details regarding these incredible advances. The collective result today is that we can routinely identify thousands of expressed genes, proteins, and metabolites even from tiny biological samples. Given time and resources, there is a good chance that we will soon have identified, quantified, and characterized a very significant portion of all types of molecules in many cell types. The Holy Grail of reductionism is emerging at the horizon.

Are all problems solved then? The answer is a resounding *No!* Just knowing all parts certainly does not mean that we understand what they do, how they play their unique roles, and what coordinates and controls them. Just imagine someone had taken the engine of a modern car apart and presented you with all its pieces. Would you be able to rebuild the engine? Probably not. But consider this: While a car engine consists of a lot of parts, a cell easily has tens of thousands of times as many, and a lot of them we do not even know yet!

So, biology is complicated, and we cannot even make reliable predictions of how a lowly bacterium will respond if we put it into a new environment. Why exactly is that? There are many reasons, which certainly are associated with the sheer numbers of different molecules, but also with their chemical features and physical structures, with uncounted processes running simultaneously in all cells, and with the complexity of interactions between all these components. This realization of complexity and unreliable predictability leads us directly into the realm of systems biology. Systems biology has begun to address these issues with new combinations of methods, including mathematics, computing, and engineering. Systems biology is still in its infancy, and that makes it a very exciting area of research. We have taken but the first steps of a long and very exhilarating path that is opening up in front of us. And we can only dream where this path may lead us.

Chapter 2
Exciting new puzzles

Systems biologists want to understand how biological systems operate within their natural surroundings. These 'systems' may be whole cells, organisms, or even populations, but they are more often comprised of biological molecules and their interactions. Because even seemingly simple systems in biology are in truth complicated, it is no surprise that investigating them together with their surroundings often means that an enormous amount of different types of information is involved. And as one might expect, this information is almost never complete. In fact, the typical situation is a collection of important core data, interspersed in a sea of gaping holes.

Systems biology has two closely interacting branches, and they address this issue in different, complementary ways. *Experimental systems biologists* use many different types of laboratory techniques to fill some of the holes with new measurements, while *computational systems biologists*, or *systems modelers* for short, depend on mathematics and computing to *infer* what is most likely happening in the holes. We will take a detailed look at both approaches in the next chapters, but because the general concepts of experimentation are relatively intuitive, I'll outline here why we need computational approaches. The answer, in one word, is 'complexity.'

The challenge of complexity

The overall task of the systems modeler is to extract information from available data and then to piece this information together, thereby generating genuinely new insights and narrowing gaps in knowledge. In many ways, one could compare the process with a huge Sudoku, where given numbers in certain locations permit true inferences regarding the missing numbers, if one just works on the puzzle hard enough. The big difference between Sudoku and computational systems biology (CSB) is that systems modelers usually don't even know whether the available information is actually sufficient to solve the puzzle, or at least parts of it. Then again, discovering even a partial solution for the first time can be exhilarating, and not knowing beforehand is therefore tantalizing and very exciting for true problem solvers. Moreover, even if one does not find a comprehensive solution, there can still be a meaningful advance if the computational analysis can identify which specific pieces of missing information would be most helpful in propelling the field forward, either by designing new experiments or collecting data from the literature or appropriate databases. The overriding challenge of this endeavor is the complexity of biological systems. I will not venture into trying to give a precise definition of complexity, but the following illustrations will hopefully create a good impression of what it is and why it challenges our minds.

Two situations are arguably of greatest interest to the systems modeler. The first situation deals with complexity: a system is so complicated that we cannot wrap our heads around it and it is therefore impossible to make reliable predictions on how this system would respond if something internal or in its surroundings were changed. The second situation consists of confusing or counterintuitive observations that traditional biology cannot explain. This situation can be found not only in large systems, but also in rather small ones.

A simple, yet true case study may illustrate this situation. The star of the story is the bacterium *Lactococcus lactis*, which is used in yoghurts, cheeses, fermented products, and other applications. *Lactococcus* is actually so important for the dairy industry that the US State of Wisconsin declared it in 2010 its State Bacterium, the first of its kind. The favorite food of this bacterium is the sugar glucose, which it takes up for energy and converts into lactate, an important ingredient in cheese. The conversion of glucose into lactate is the result of several biochemical reactions that collectively form a *pathway* called *glycolysis*, which translates into 'the break-down of glucose.' Humans also employ glycolysis for energy production, using the common molecule ATP (adenosine triphosphate) for the first reaction. *Lactococcus* instead uses a compound called PEP, which is short for phosphoenolpyruvate. The interesting aspect for our case study is that PEP itself is a molecule that is generated during glycolysis. Somewhat simplified, the pathway of glycolysis in *Lactococcus* is depicted in Figure 1(a). The bacterium takes up glucose by moving it across the cell membrane and immediately turning it into compounds A and D, using PEP in the process. Afterwards, A is converted into B, then C, which becomes PEP, from which D is produced. D is converted into lactate, which leaves the cell. So, that looks easy enough.

Now the puzzle begins. Imagine that glucose runs out, which happens frequently in nature. The obvious consequence is that no more A, B, C, D, and PEP can be produced, and all material flows through the system like through a chain of pipes and ends up as lactate. Now suppose that, after a while, glucose becomes available again, and *Lactococcus* would of course like to take it up, but it has a dilemma: all PEP is used up, and without PEP, no A can be produced. Ever again. If you don't think that sounds right, you are absolutely correct: After all, the bacterium has been around for a very long time. So, what's missing?

The way a systems modeler approaches the issue is to think about possible mechanisms with which the bacterium could possibly

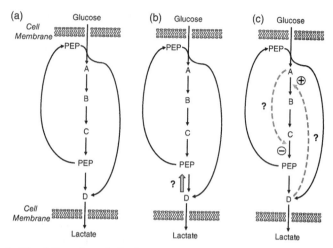

1. Glycolytic pathway in the dairy bacterium *Lactococcus lactis*.
(a) The bacterium takes up glucose from its surroundings across its cell membrane, and with the help of the compound PEP, converts it ultimately into lactate. (b) Could it be that there is a reverse reaction from D to PEP? (c) Would an inhibition signal from A or an activation signal from D solve the puzzle discussed in the text?

Systems Biology

remedy the situation. Any promising possibility is formulated as a specific hypothesis and converted into a computational model, which typically consists of specifically chosen equations. We will discuss this process in Chapter 4 in greater detail. Now the modeler performs mathematical and computational tests with this model. Together, these tests are designed to show whether the hypothesis, if true, permits *Lactococcus* to take up glucose again. If the model confirms that the hypothesis is a possible solution to the problem, the systems modeler presents the hypothesis to an experienced microbiologist for validation. If not, the search goes on for better hypotheses or for additional data.

For example, one could think that some of the compound D could be converted back into PEP (Figure 1(b)). The argument could be that, while the diagram in Figure 1(a) does not exhibit this option, it could be conceivable that this reaction was never reported, even though it exists. In this case, a model analysis would indeed confirm this mechanism as a possible solution, if one could prevent D from leaving the cell. Caution, modelers! You should always discuss theoretical solutions with biologists. In this case they would counter that targeted experiments demonstrated very clearly that this back-reaction was never detected in *Lactococcus*. So, a model 'solution' may be a mental victory, but if it does not answer the biological question, the search continues.

What if one allows *regulation* of a type where the existence of one of the compounds either activates or shuts down one of the reactions? These types of inhibition and activation signals are found all over biology, which implies that hypotheses involving some sort of regulation could indeed be true. As an example, compound A could possibly *inhibit* the reaction between C and PEP (Figure 1(c)). So, if a lot of A is present, not much C (or no C at all) is converted into PEP. As a different example of regulation, suppose D would *activate* the reaction between A and B (Figure 1(c)). If so, A would not be converted into B unless D was present. Each of these speculations is tested with a model and must then be validated by a biological experiment. As a mental exercise, think about types of regulation that would allow *Lactococcus* to restart glycolysis if glucose were depleted for a while but then became available again. To be specific, this regulation would require that one or more of the compounds (glucose, A, B, C, D, PEP, pyruvate, or lactate) would either activate or inhibit one or more of the reactions, which are represented by arrows. *Lactococcus'* own solution, using regulation, is given at the end of the chapter.

There is no limit to the sophistication of puzzles found in nature. For instance, we know a lot about chronic diseases, but we have no definite explanation for why some people get cancer or depression and others do not. Predictions regarding the next flu strain have improved a lot, but they are still wrong once in a while. Why? And when it comes to depression, we are not even close to figuring out how simple nerves can interact with each other to create thoughts or memories. Even simple organisms can be puzzling: we have amassed a lot of information about the responses of microbes to various stresses, such as heat or pressure, but we do not fully understand how these responses are coordinated. And the long list goes on.

What exactly makes these phenomena so difficult to explain? One immediate culprit seems to be the size of a system. Very large systems, with many components, processes, and signals, can be overwhelmingly complicated. For instance, the human brain allegedly contains at the order of 150 trillion connections between nerves, called *synapses*. So, it's no wonder we have problems understanding the brain! But size alone cannot be blamed. As an illustration, imagine a very long chain of processes, such as a hundred domino tiles neatly arranged, all standing on their sides next to each other and with short spaces between them. Barring failure due to outside forces or clumsiness, we can predict very accurately what will happen if the first domino is tipped over toward the next one. With some timing experiments, we can even get a pretty good estimate for how long it will take before the last domino tumbles. Amazingly, or not, the same is true for 10,000 or 10,000,000 dominos.

All domino tiles in this chain are the same. Is that the reason for the ease with which we can predict what happens? Maybe partially, but not really. Consider the 1931 Rube Goldberg concoction of a self-operating napkin that the fictitious Professor Lucifer Gorgonzola Butts, A.K., invented and that even made it onto a US postage stamp (Figure 2). (As an aside, nobody seems to be sure what 'A.K.' stands for; possibly 'all knowing'?)

Goldberg himself explained the function of the self-operating napkin 'system' in the 26 September 1931 issue of *Collier's Weekly*:

2. **Professor Butts's invention of a self-operating napkin, according to Rube Goldberg. Trace and try to understand the proposed sequence of events from A to N or read in the text how Goldberg himself explained the function of the 'system.'**

As you raise spoon of soup (A) to your mouth it pulls string (B), thereby jerking ladle (C) which throws cracker (D) past parrot (E). Parrot jumps after cracker and perch (F) tilts, upsetting seeds (G) into pail (H). Extra weight in pail pulls cord (I) which opens and lights automatic cigar lighter (J), setting off sky-rocket (K) which causes sickle (L) to cut string (M) and allow pendulum with attached napkin to swing back and forth thereby wiping off your chin (N). After the meal, substitute a harmonica for the napkin and you'll be able to entertain the guests with a little music.

Goldberg's 'system' is quite interesting, for several reasons. First, the diversity of components does not play tricks with our ability to predict the ultimate outcome: if everything works as intended, it is easy to foresee what will happen if one raises the spoon, because the entire process is a linear chain of events that happen one after another. Admittedly, the in-between steps are a bit outlandish but, barring unforeseen mishaps, the napkin machine will ultimately wipe the chin. Second, the process of reaching N from A appears to be much too complicated, and one has the impression that there should be simpler solutions. However, cells and organisms also often use processes that could potentially be simpler, but this complexity is at least partially due to the discrete forward steps of evolution and the constraints these steps had to satisfy at every point in time. Third, and arguably most important, despite the complexity of Goldberg's machine, there is no redundancy: if one part or process does not operate perfectly, the entire system fails. By contrast, nature almost always has a 'Plan B.' Sometimes we do not understand the complexity that enables a particular Plan B and might think that nature is unnecessarily complicated. But studying these complexities we often find that they compensate so smoothly for failures in other components or processes that we don't even notice these failures. Finally, Goldberg's system can do its job only once before it needs to be set up again, unlike *Lactococcus*, which can stop and go indefinitely.

As a stark contrast to this complicated-looking, but in truth simple, napkin system, let's look at the simple-looking, but in truth confusing, system in Figure 3, which consists of only two components, A and B. In a biological context, one might think of

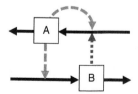

3. Simple-looking two-component system with confusing responses. A and B activate (dashed arrows) or inhibit (dotted arrow) production and utilization processes. A realistic example could be a pair of genes that affect each other's transcription.

two genes affecting each other's expression, but the example need not be biological. In this little system, A and B are both produced and utilized (solid arrows); A activates the production of B, as well as its own production (dashed arrows); and B inhibits the production of A (dotted arrow). Suppose both A and B are in a *steady state*, which means that they are not changing in quantity or concentration because their production and utilization perfectly balance each other. Now let's try to predict what happens if we increase A, let's say by 20 per cent. Tracing the arrows, we see that the rise in A leads to increases in the two activation signals (dashed arrows), which trigger increases in the production of both A and B. More B means more inhibition of the production of A, which now fights the activation of this same process by A. Who will win? Will A go up or down? And what about B? Will it go up or down?

The answer to these questions is disturbing: it is impossible to predict the outcome, unless we set up a mathematical model with actual numbers that specify the magnitudes of production and utilization and the strengths of all activation and inhibition signals. Even if we are given these numbers, our brains are not capable of making reliable predictions, and only a mathematical model can provide true answers. The reason is that, depending on the specified numbers, very different things may happen. Both A and B may keep on growing without end. For other numbers, both A and B decrease to 0, either immediately, or after some ups and downs. For other numbers, both A and B may oscillate for a while and then return to the steady state. Finally, for yet other constellations of numbers, both A and B may start to oscillate and keep on oscillating until some outside force puts an end to it. This *emergence* of new behavior is often observed in biological systems but is really hard to explain. Here, for instance, neither A nor B oscillates by itself, so where do the oscillations come from? The sobering conclusion is that our brain by itself is insufficient to solve the puzzle, and even hard thinking is not enough.

The numerical features of the system divide the realm of all possible responses into distinct sections, within which the system 'blows up' (the variables grow without end), 'dies' (the variables go to zero), or where one finds different types of oscillations ('damped' or 'stable'). These sections are separated by invisible thresholds or dividing lines that can only be identified with rigorous mathematics. The important insight from this example is the demonstration that even very small systems can throw off our intuition. And it is not even the case that we could practice our intuition for these situations, because the different system behaviors may depend on the third (or one-hundredth) numerical value behind the comma of one of the quantities that define the system.

The all-overshadowing reason we cannot make predictions in this and many other cases is that this system is no longer *linear* like the domino example and Professor Butts's invention (Figure 2), where one step neatly follows the previous step. The structure of *nonlinear* systems is usually such that changes in one part eventually come around and affect this part itself. The action of this *feedback* cycle may be diminishing after a while or amply itself. In addition, there is often competition between activation and inhibition, which our brain is unable to evaluate without mathematics, or there are invisible points or lines where the system all of a sudden comes up with a new and unanticipated response. To put it simply, 'linear' means straight and easy to foresee, whereas 'nonlinear' means curved and often unpredictable.

To gauge the effects of *nonlinearities*, let's first look at linear systems again, whose behaviors are much more intuitive than those of nonlinear systems. As an illustration, imagine a bakery that has two ovens with the combined capacity of baking eighty loaves of bread per day. If so, six ovens would have the capacity of 240 loaves. Within reasonable limits, we can easily and correctly scale these numbers up or down. By contrast, giving a chicken three times its normal feed will not result in three times as many eggs, indicating that scaling does not work well in this case. This situation is typical for all living systems. They have a lot of internal limits, and if a system bumps against such a limit, its output either does not change any further or is even reduced, or the system switches to an entirely new mode of response where it does something totally different. If we are a bit chilly, changes in brain hormones lead to constricting blood vessels. This constriction decreases blood flow and raises blood pressure in the skin, which in turn enhances the heat insulation of superficial tissues. But if that warming method is not sufficient, we start producing heat by shivering, which is an entirely different response mechanism. In addition to these immediate responses, the body slowly starts to adapt to colder temperatures after about two weeks of cold exposure.

Let's look at another example. A US penny weighs 2.5 grams and a Euro cent coin weighs 2.3 grams. Therefore, two pennies and two cents weigh 9.6 grams, and 4,000 pennies and 6,000 cent coins weigh (10,000 + 13,800) grams = 23,800 grams. The weights simply add up in a linear manner, no matter how many pennies and cent coins we have, and this feature is called *superposition*. By contrast, essentially all biological systems and processes are nonlinear, and gradually increasing the same quantity can lead to very different outcomes. For instance, a pinch of salt may improve the taste of a bowl of soup, and two pinches could possibly even further enhance the taste. However, there is clearly a limit to the improvement, and a hundred spoons of salt might just make the soup inedible: taste is a nonlinear phenomenon. Likewise, moderate sun exposure usually leads to a sun tan, but a little bit more sun causes sunburn, which is an entirely different physiological response from a tan. There is no superposition.

The missing superposition in nonlinear systems is a much bigger challenge than it might seem, because superposition allows us to analyze individual parts in separate experiments and, in the end, to 'add' all results together. That simply does not work for nonlinear systems. At the same time, nonlinear systems are much more versatile and interesting, as they can exhibit all kinds of oscillatory and chaotic responses, in addition to thresholds where the system responses suddenly change to something entirely new, such as sunburn. This flexibility makes the analysis of nonlinear systems challenging but also very exciting.

Much of the nonlinearity and unpredictability of biological systems is caused by *regulatory control mechanisms* that operate in distinctly different ways and at different time scales. A good example is the regulation of *metabolic processes*, which are biochemical reactions that convert food into energy and into a vast number of chemical compounds our body needs.

As we discussed for the glycolysis of *Lactococcus*, a first means of regulation is a metabolite modulating some reaction in the system, either by activating or inhibiting it. Specifically, most biochemical reactions are *catalyzed* by *enzymes*. These enzymes are proteins that make biochemical reactions possible. They have a three-dimensional shape with a complicated surface, and if the shape of a metabolite fits into a specific 'pocket' in this surface, it can slow down or speed up the activity of this enzyme, thereby inhibiting or activating the reaction. A metabolite may regulate different reactions, and the same reaction may be regulated, up or down, by different metabolites.

The regulation usually targets branch points in a metabolic pathway, where a compound can be used for different purposes. For instance, compound A in Figure 4 may be used for the production of both B and C, which could be the starting points for other chains of processes. This type of branched system is ubiquitous throughout metabolism, and also in many other applications where a component can ultimately have different fates. An example for A is the amino acid serine, which can be converted by different enzymes into the amino acids glycine and cysteine. Without regulation, the material from A would always be

20

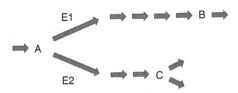

4. Branched metabolic pathway, in which A can be used for the production of B and C. The two branches are controlled by the enzymes E1 and E2, and activating or inhibiting these enzymes allows the cell to channel more of A toward B or toward C.

divided up in some fixed proportion to become B or C. But if the cell needs to control how much B or C is required at some time point, for instance in order to respond to a change in its surroundings, it needs to be able to alter this proportion. A cell can accomplish this task, among other options, by inhibiting enzyme E1. This inhibition can possibly be exerted by a number of metabolites, but in particular by B itself, so that production is reduced if a lot of B is already there. This *feedback inhibition* secondarily results in more of A flowing towards C than before. The cell could also activate E2, alone or in combination with inhibiting E1, to increase production of C.

As a second mode of regulation, a cell can control the magnitude of a reaction by activating or deactivating its enzyme. This adjustment is accomplished by other enzymes that attach to—or detach from—the enzyme specific molecules, such as phosphate groups. These mechanisms are quite fast.

Furthermore, all proteins are continuously recycled and their abundance is regulated through the balance between production and removal. The cell makes use of this fact by degrading an enzyme if it is no longer needed in the current amount. Specific enzymes are available for this protein disassembly.

Finally, if a greater amount of some metabolite is needed on a regular basis, the cell can upregulate the genes that code for those enzymes catalyzing the reactions involved with the production of the needed metabolite. As a variation, the cell may regulate the corresponding transcription and translation processes, from DNA (deoxyribonucleic acid) to RNA (ribonucleic acid) to protein, which ultimately lead to more of the enzyme and thereby increased production of the desired metabolite. For instance, if the amount of E1 is increased in the pathway of Figure 4, the former ratio of material flowing toward either B or C is shifted toward B. This mode of transcriptional and translational control is slower and also longer lasting than regulation by a metabolite or disassembly of a protein.

In a way, the different modes of regulation serve the same purpose, namely to channel more material into pathways of high demand and decrease material flow into pathways that are temporarily needed less. However, they differ in their details and thereby help a cell control its metabolism with different speeds and for shorter or longer periods. Indeed, all modes of regulation may be called up at the same time.

Throughout the millennia, evolution has hard-wired additional control strategies into cells. For instance, yeasts produce a double-sugar called trehalose, which protects cellular structures against heat and other adverse conditions. Trehalose is produced by two enzymes, E1 and E2, and degraded by one, E3 (Figure 5).

5. **Trehalose response to heat in yeast. Close to the optimal temperature for baker's yeast, *Saccharomyces cerevisiae*, the trehalose production enzymes E1 and E2 have relatively low activity, whereas E3 has high activity. As a result, the cell does not contain much trehalose. By contrast, if the temperature rises to about 35°C, the pattern flips, with high activity for E1 and E2 and low activity for E3. With this setting, trehalose is available in large quantities.**

Interestingly, these enzymes have their highest activity for different temperatures: at the yeast's 'favorite' temperature of about 25°C, E1 and E2 have relatively low activity, whereas E3 has high activity. As a result, not much trehalose is produced. By contrast, if the temperature rises to about 35°C, the pattern flips, and trehalose is produced in large quantities and not removed. These settings are not controls that the cell can call up in times of need. Instead, the enzymes have evolved in such a manner that their physico-chemical properties 'automatically' switch from low to high trehalose production if the temperature rises. Moreover, production and degradation in this system are occurring all the time, which may appear to be a waste of energy. However, this set-up allows an almost immediate switch to high—and back to low—trehalose production when the temperature rises and when it returns back to normal. It appears that these fast responses offer the cell a selective advantage, thus favoring this set-up throughout evolution.

One type of regulation is at first counterintuitive, because the same 'Source' component sends both an activating and an inhibiting signal to the exact same 'Receiver' process. This design is found in the transmission of signals within a cell or organism and called an *inconsistent pathway*. Such a pathway has at least two parallel branches (Figure 6). According to the upper branch, a change in Source activates the Receiver, but tracing the lower branch, we find that the same change in Source inhibits the Receiver. Can this internal competition possibly make sense? Yes, if the two signals require different lengths of time to reach the target. In the given example, the one-step activation is essentially immediate, whereas the inhibition takes a series of steps. As a consequence, this design allows the cell to use a change in Source to turn the Receiver process on for a short period of time and then to turn it off. An example of this type of signaling mechanism—whose molecular details are, however, much more complicated—is a system of molecular pathways in plant leaves that opens and closes openings for gas exchange, called stomata, and is responsible for the plant's response to drought; it is discussed in Chapter 5 (Figure 13).

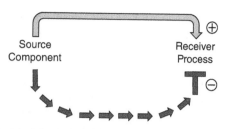

6. Inconsistent pathway. Any change in the Source component is transmitted in one step (hollow arrow) to activate a Receiver process. The same change is also transmitted via the darker arrows, but more slowly, ultimately turning off the process. At first, this 'internal fight' seems to make no sense. However, it is useful for activating a process for just a short period of time and then resetting it to its normal level.

Where to start?

There is little doubt that life is complicated. We saw that even small systems and their behaviors can evade our intuition. If we consider the complexity of realistic biological systems, their many components and different levels or organization, their nonlinear processes, and their intricate webs of regulation, we might find the task of systems biology overwhelming. Yes, it is true: we are faced with grand challenges. But keep in mind that systems biology is barely in its infancy and that the complexity of nature offers ambitious researchers truly tantalizing puzzles with potentially huge rewards and worldwide implications. So, let's take the first step on the road before us and see what's out there in terms of data and other information that we need to solve problems with models. The next chapter will be reassuring. It demonstrates that recent advances in molecular biology and other fields have been creating incredible molecular inventories. Just a few decades ago, these comprehensive catalogues of the components of cells and organisms seemed absolutely impossible to obtain. Now they are available, and they expand every day. But before we delve into this new realm of big biomedical data, we have one piece of unfinished business.

How *Lactococcus* does it

In the glycolysis diagram for *Lactococcus* (Figure 1), the compounds A and B have the chemical names glucose 6-phosphate (G6P) and fructose 1,6-bisphosphate. Importantly, but not shown before, they activate (dashed arrows) the enzyme pyruvate kinase (PK), which controls the conversion of PEP into D (Figure 7), which is chemically called pyruvate. If A and B are absent, PK

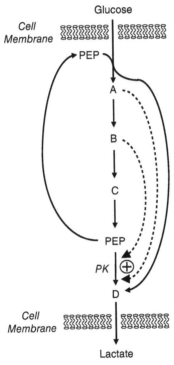

7. **Regulation of glycolytic pathway in *Lactococcus*. In the actual bacterium, the compounds A and B activate the enzyme controlling the pyruvate kinase (PK) reaction between PEP and D. This regulation makes the start-and-stop mechanism of glycolysis possible.**

stops catalyzing this conversion. Suppose that glucose runs out. Shortly after, A and B are used up, which shuts down the activity of PK, so that no more PEP is converted into D. The important consequence is that a modest amount of PEP is left. Now the bacterium waits in this state of suspended animation. As soon as glucose becomes available again, the remaining PEP permits the conversion of glucose into A, and glycolysis starts again, with the help of renewed PK activation. A computational model confirms the effectiveness of this stop-and-start mechanism. In reality, the solution is more complicated, but the key mechanism is this activation. Theoretically, it could also be C that activates PK.

Chapter 3
The —omics revolution

Sherlock Holmes once famously said 'It is a capital mistake to theorize before one has data.' Philosophy and pure mathematics may do just fine without data, but in biology the verdict is clear: without data we can only speculate rather vaguely; all progress relies on good data. In the past, experiments were time consuming and expensive, and data were therefore often scarce. The so-called —*omics revolution* has changed that to a point where we often have so many data that we cannot make sense of them and need to resort to sophisticated computing methods. This shift, which is the topic of this chapter, even changed how we perform experiments and think about science.

Biology has been around for a long time and, over the centuries, biological data have changed tremendously. Initially, observations were qualitative. A typical observation might have been 'Plants grow taller if enough nitrogen and phosphorus are available.' Indeed, even today, qualitative information can be very useful. For instance, oysters and mussels are considered *indicator species* or *biomonitors* in the sense that their mere existence provides information about the health of coastal waters. Over time, qualitative statements were made more precise with appropriate adjectives. But, while an improvement, that is not enough. The philosopher of science Rudolph Carnap lucidly explained why science really needs quantitative terminology:

> First of all—though this is only a minor advantage—there is an increase in the efficiency of our vocabulary....Without the concept of temperature, for example, we have to speak of something as 'very hot,' 'hot,' 'warm,' 'lukewarm,' 'cool,' 'cold,' 'very cold,' and so on...What would be wrong with this? For one thing, it would be exceedingly hard on our memory. We would not only have to know a great number of different adjectives, but we would also have to memorize their order, so we would know immediately whether a certain term was higher or lower on the scale than another. But, if we introduce the single concept of temperature, which correlates the states of a body with numbers, we have only one term to memorize. It is true, of course, that we must have previously memorized the numbers, but once we have done so, we can apply those numbers to any quantitative magnitude....The major advantage...is that quantitative concepts permit us to formulate quantitative laws....Once we have the law in numerical form, we can employ that powerful part of deductive logic we call mathematics and, in that way, make predictions.

Biology does not have such laws yet (see Chapter 7), but it has become evident that progress would be impossible without rigorous methods that generate quantitative data consisting of numerical values and units.

By and large, every data point is the result of an observation or experiment. Over the centuries, legions of biologists have been describing new species, originally in Latin, based on counting, measuring, and weighing their features. This tradition continued seamlessly into the quantitative characterization of organs, tissue, cells, and the components of cells, collectively leading to an incredible inventory of biological structures and their details. This quest of determining 'what is there' is going on today and will continue for a long time. In contrast to the past, most of this quest now targets the structure and location of molecules like proteins and lipids rather than the identification of new species.

If 'what is there' is one side of the coin, then 'what does it do' is the other. Both go hand in hand, but the latter is much more difficult to investigate than the former, because it contains aspects of time and causality, which involve direct and indirect processes that can hardly be assessed with snapshots. For instance, a picture of a human cell, obtained with electron microscopy, clearly shows mitochondria and other cellular structures, but the picture cannot tell us that these mitochondria are the power plants of the cell and, thus, the entire organism. How can we find that out?

A good strategy is the application of the scientific method, whose core is the testing of hypotheses. In the case of mitochondria, a hypothesis could be that their role is the generation of ATP, which the cell can use for energy; this hypothesis has been shown to be true. Ignoring experimental details, a test could consist of killing all mitochondria in a cell and measuring ATP. A wrong hypothesis could be that mitochondria are vehicles for transporting materials in and out of the cell. An experiment could look for mitochondria outside the cell, but would not find any, thus refuting the hypothesis.

Numerous variations of the scientific method can be found in the literature and on the Internet. A typical formulation is given in Table 1. The central concept though is always a well-posed, testable hypothesis. Thinking up and testing such a hypothesis requires background knowledge and good intuition regarding the observation of interest. Almost all real progress in biology has traditionally come from this process of formulating, testing, confirming, or refuting hypotheses.

In stark contrast to this tried-and-true procedure, a new way of doing biology has been emerging since the early 2000s, and it proposes a very different strategy. This new approach is directly tied to the —omics revolution. The terminology is quite odd,

Table 1. The traditional scientific method of biology

Step	Action
1	Make an observation
2	Ask a question regarding the observation
3	Do background research to see what is known about the question
4	Refine the question to a point that it can be formulated as a hypothesis
5	Design an experiment to test the hypothesis and execute it
6	Analyze data resulting from the experiment and determine to what degree the results support the hypothesis
7	If the results do not support the hypothesis, go back to 2, 3, or 4
8	If the experimental results do support the hypothesis, new insights are gained. Share these insights with the scientific community
9	Usually new insights lead to additional questions. Formulate new hypotheses that target these questions

as '—omics' is not really a word, but rather a suffix that has been attached to various biological terms to indicate 'all of it' or 'a lot of it.' The best-known example is *genomics*, which means studying the structure, function, and evolution of genes; not one gene at a time, as it used to be done a few decades ago, but for all (or most) genes at the same time. Similarly, *proteomics* deals with aspects of many or all proteins, and *metabolomics* addresses many or all metabolites in a biological sample.

The possibility of studying very many genes, proteins, or metabolites simultaneously has been the result of innovative collaborations between biologists and engineers. In particular, these approaches became practical and feasible due to a combination of better measurement techniques, miniaturization, robotics, and computational data analysis, which now permit molecular assessments even in very tiny samples. A good example is a so-called DNA microarray, which is a silicon chip on which robots have placed tiny amounts of DNA and which allows the researcher to measure—in a single experiment—to what degree thousands of genes are turned on or off.

The —omics revolution has not only generated huge datasets, it has turned the tried-and-true scientific method on its head. The central position of a strong hypothesis has vanished, and the new mindset is *exploratory data analysis*, which here translates into the lax mandate: 'let's see what's out there and then we'll try to figure out what it means.'

A good example is a gene expression based approach toward cancer. Suppose the magnitudes of expression of all genes in a healthy cell and of all genes in a corresponding tumor cell were measured and compared. In contrast to the traditional scientific method, there is no hypothesis that gene X or gene Y might have something to do with that particular cancer, but rather all genes are examined in the hope of detecting *patterns* (see Chapter 4). Substantial differences in expression between corresponding genes in healthy and cancer cells suggest that these genes might be involved with cancer growth. Maybe surprisingly, it is often not even known what the specific roles of these genes are, but that is beside the point: the prime target is any *significant difference* in expression. Of course, this difference could be totally independent of the tumor, but if the same differences show up again and again for different patients, and maybe in different types of cancer, it is likely that there is some truth to the discovered pattern. Now the next step is to determine what the pattern means.

The same holds for proteomics and metabolomics. In the latter, for instance, urine samples are taken both from a healthy person and from a person with kidney disease and, using the method of mass spectrometry, the amounts of hundreds or even thousands of metabolites are quantified. A comparison between the profiles may identify differences that lead to specific hypotheses regarding the disease.

The new scientific method of experimental systems biology is summarized in Table 2. In the first step, an interesting question within a pertinent biological context is identified. As an example, suppose we are interested in identifying *molecular signatures* associated with liver cancer. Such signatures are noticeably elevated or reduced concentrations of some metabolites or proteins, or significant changes in the expression of particular genes. Depending on our prior knowledge of the disease, we need to decide whether the expected changes are most likely associated with genes, proteins, metabolites, or other classes of molecules. Suppose we have biological reasons to believe that the expression of some genes is affected in liver cancer and that the changes depend on the age of the patient. It is now necessary to obtain

Table 2. The new scientific method of experimental systems biology, applied to a disease

Step	Action
1	Identify an interesting phenomenon, such as a disease
2	Decide whether transcriptomics, proteomics, or metabolomics is most promising
3	Perform the same —omics analysis of genes, proteins, and/or metabolites with a sample from a healthy person and from a person with the disease
4	Collect significant differences between corresponding measurements in the two samples, using methods of machine learning
5	Assemble these differences into normal and disease profiles
6	Formulate hypotheses explaining the differences between these profiles
7	Perform traditional experiments testing the most promising hypotheses
8	If the results do not support the hypotheses, go back to 2, 3, or 6
9	If the experimental results do support one of the hypotheses, new insights are gained and should be shared with the scientific community

biological samples and corresponding controls from liver cancer cells and from corresponding healthy cells, in both cases from people of different ages.

The samples are used for a *transcriptomic* study that collectively quantifies to what degree the genes in all samples are turned on or off. The result is a large dataset consisting of the expression of most or all genes in healthy and liver cancer cells of men and women of different ages. The results could also consist of a combination of gene expression and the amounts of proteins and/or metabolites. Because the resulting datasets consist of thousands of data points, they are evaluated with computational methods of statistical machine learning (ML; see Chapter 4) which, in a successful study, reveal *patterns in the data*. For instance, we might find that the expression of a small number of genes is always much higher in cancer cells than in healthy cells. In an ideal situation, these genes would have known roles. They could be associated with metabolic pathways, signaling pathways, or some specific physiological mechanism, such as a response of the immune system. In less than ideal situations, the patterns are unclear and/or the specific roles of the affected genes are unknown. It is often not even known whether specific genes code for proteins or whether they serve other purposes, such as coding for RNAs that regulate the expression of other genes. In opportune cases, the results can be reformulated as specific hypotheses, such as: liver cancer cells exhibit reduced mitochondrial function, and the consequences are stronger in older patients. Such a crisp hypothesis can be tested with targeted experiments that assess mitochondrial function. If the experiments confirm the hypothesis, new insights are gained. If they do not, we must go back and analyze the data with different methods or collect additional data.

The differences between the traditional and the new scientific method are driven by the availability of data and by effective methods of analysis. In contrast to even the late 20th century, data are often cheap now, and computers can sift through millions of data points in order to find needles in haystacks (see Chapter 4). As an immediate consequence, it has become feasible to address numerous possible explanations of a phenomenon, by generating very large datasets. Interestingly, this process actually ignores most of the results; for instance, it doesn't consider all genes that show the same expression patterns, and in the process it filters out relatively few significant differences that point toward new, testable hypotheses. Of greatest importance is that the research process no longer begins with a specific hypothesis, but rather with the automated extraction of relatively small amounts of significant information from huge datasets. While most of the acquired information is not explicitly used, the results of this filtering often suggest the creation of new hypotheses, which would not have been envisioned otherwise.

Chapter 4

Computational systems biology

The new methods of —omics biology, combined with more traditional experiments, have the capacity of generating more high-quality data than ever before. So, why isn't that sufficient? What is missing? The missing aspects arise from subtle but important differences between *data, information, knowledge,* and *understanding.* These differences have been discussed by philosophers and information scientists for many decades, but it is sufficient here to reduce their insights to a crude summary. *Data* are discrete, objective facts, often without a context. Let's say you heard 'rainfall at the airport yesterday was 2.6 centimeters.' Suppose that this is a true fact; but what does it really mean? Is 2.6 centimeters a lot? Is it a normal amount? Was the airport flooded? Were flights cancelled? Does something need to be done about rainfalls of this magnitude in the future? We can easily see that the singular data point is not all that telling or interesting unless we know much more about the background against which it should be gauged. For an airport in Saudi Arabia, 2.6 centimeters of rain in June is highly unusual, whereas the same amount is not even worth mentioning during monsoon season in Bangladesh. In the example, location and timing augment the data point with meaning. Generally, *information* consists of processed data augmented with *metadata*, which consist of such meaning, together with some context and background. Moreover, *scientific information* often has the connotation of being something useful or relevant, something that has a purpose. *Knowledge* is more difficult to define in unambiguous terms. It is a mixture of synthesized information, context, experience, and possibly even value judgment. It explains how the information was obtained or generated, recognizes patterns in information, offers procedures for assessing a situation (*know-how*), and guides expectations regarding the future. *Understanding* interprets knowledge. It provides causality and rationale for a phenomenon, detail, or pattern in data. Understanding helps us explain why something is the way it is.

Understanding living systems is the ultimate target of biological science. But laboratory experiments generate data, whereas understanding additionally requires significant human intelligence and knowledge. Computational systems biology (CSB) attempts to bridge the gap between data and understanding.

For small datasets, an experienced biologist usually has good intuition about their meaning, and there is often plenty of other information that serves as context. However, if the data come in large and complex datasets, it can become difficult to interpret them and extract valuable information. For instance, a microarray experiment may reveal differences in the expression of over 20,000 genes in normal cells and cancer cells of the same tissue of the same person. Of course, we can look at one gene at a time and may find a difference in expression of, say, 15 per cent. Is the cancer cell *significantly* different? Is the 15 per cent difference coincidence? Is it worrisome? It is impossible to say. A better and more important question is: what does the entire *profile* of differences mean? CSB offers assistance with the interpretation of complicated biomedical data of this type.

Painted in very broad strokes, CSB uses a pipeline from data to understanding that consists of two toolsets, which are discussed in this chapter. The first consists of computational methods of *machine learning* (ML) that extract as much true information as

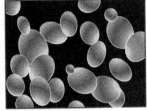

**8. Reality and model. Left panel: Yeast cells of the species
Saccharomyces cerevisiae NEU2011. Right panel: Assembly of
ellipsoids with different lengths and ratios of axes. For many purposes,
the similarity between the actual cells and the mathematical
representation is probably sufficient.**

possible from large sets of raw data, while filtering out spurious
results and errors in the datasets. The second toolset overlaps a bit
with these methods. It analyzes *mathematical models*, which span
the spectrum from very simple to extremely complicated. As a
simple example, the shapes of yeast cells may be *approximated*
with ellipsoids (Figure 8). The term 'approximated' means that the
representation of a yeast cell as an ellipsoid is not entirely true,
but that it is simpler and yet accurate enough for the questions we
try to address. As an illustration, the ellipsoid approximation
allows us to estimate the total volume of the yeast culture, just by
counting cells and using the volume formula of an ellipsoid. For a
rough estimate, we may even just use the size of a cell that 'looks'
average, but we can of course improve the estimate by measuring
lengths and widths, which is still easier than measuring volume.

The more complicated models fall into two categories. The first
category contains *static networks*. As an analogy for this concept,
imagine the roadmap of a country. In network jargon, the cities
and villages are generically called *nodes* and roads directly
connecting them are called *edges*. Although the total number of
people in each city, including visitors and people travelling
through, is likely different from day to day, the map itself does not
change all that much, at least not on a daily basis; it is *static*.

Typical nodes of networks in molecular biology are proteins, and an edge between two proteins may indicate that they attach to each other.

The second category of models is an extension of networks to *dynamic biological systems*. Not only may their nodes change over time in their magnitudes, their edges may be controlled by some of the nodes, so that they allow different amounts of traffic, mass flow, or information flow, depending on demand. A typical example is metabolism, where the amounts of many compounds change after a meal and the biochemical reactions between them are regulated, for instance, by inhibitors, as we discussed in Chapter 2. Because it is more difficult to represent and understand changes in systems, current analyses usually target systems that are much smaller than static networks. The methods of this toolset create and analyze *dynamic models* that contain all important aspects of an actual biological system in the form of mathematical representations or computer programs. As an example, the temperature of the human body changes during the day and night cycle, but although it is affected by all kinds of factors, it may be approximated quite nicely by a simple sine function (Figure 9). The advantage of translating biology into a model is that this model permits all kinds of analyses, as they are discussed in this chapter.

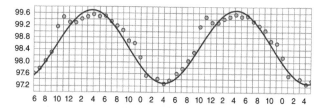

9. **Human core body temperature in °F, beginning at 6:00 am and, for illustration purposes, copied for a second day (symbols). Superimposed is a sine function. It does not fit perfectly, but is a simple and effective approximation that may be good enough for certain analyses.**

Machine learning

ML is a scientific field that studies datasets by combining statistics with heavy-duty computing. It has a lot of overlap with *data mining* and *big data analysis*. The overarching goal of ML is to train computer programs (*algorithms*) to find something in a dataset that had not been known before, such as a nuanced association between two types of data (see below). Of course, ML is not limited to biological data. We encounter it every day in many societal contexts. For instance, information gleaned from big datasets can be of great interest to pollsters advising politicians or to merchants who attempt to find patterns in their customers' spending behavior. In the latter case, every credit card transaction is stored somewhere, and due to the enormous number of such transactions, computer algorithms can be trained to detect statistically robust patterns, which humans can then interpret and act upon.

ML falls into the field of artificial intelligence (AI) but has a much more limited scope than AI. Although its results can be amazing, and the accuracy of findings steadily improves as more data become available, extracting associations from datasets is a comparatively straightforward computational process. By contrast, AI is much more complicated as it mimics human behavior in an attempt to acquire knowledge. This knowledge is used to make decisions or solve problems that normally require human intelligence, such as recognizing a face or voice, or translating text from one language into another.

ML requires training that can happen with or without human supervision. In the former case, the human supervisor gives the algorithm a large dataset, let's say, containing data from healthy individuals and from individuals with lung cancer, informs the computer which individuals have cancer and which do not, and instructs the algorithm to find features or patterns in the dataset that distinguish healthy from diseased individuals. For instance, the algorithm might find that lung cancer is often, but certainly not always, associated with cigarette smoking or with the upregulation of certain genes. If the dataset is large and representative of many variations of health and disease, the algorithm can possibly detect all kinds of data profiles in individuals with lung cancer and forge them into patterns that could be tell-tale signs of lung cancer. Once the algorithm is trained, it is given new datasets and asked to predict which individuals are at an elevated risk of lung cancer, based on the patterns it had learned during training. In many cases, this type of ML is used to classify large datasets generically into *normal* and *abnormal*.

The algorithm may also learn without human supervision. In this case, it does not know which individuals actually have cancer and which do not. Instead, the algorithm is asked to determine whether there are *any* significant patterns in the dataset. These so-far unknown patterns typically consist of associations among certain groups of data. For instance, the algorithm may detect that entire groups of genes are frequently up-regulated together or down-regulated together, depending on the conditions under which gene expression was measured. If so, these groups may be governed by the same control mechanisms or respond similarly to certain stresses or diseases. Under favorable conditions, this type of algorithm suggests new hypotheses that are to be interpreted and tested.

Because ML uses computers, large numbers of data are no real problem. Millions of data points are easily organized, stored, analyzed, and filtered according to a wide range of criteria the researcher can specify for the given case. Due to this flexibility, ML has become commonplace for many types of —omics analyses, but while its basic concepts are relatively easy to understand, there are, of course, many unsolved challenges associated with the technical details of these methods. ML is therefore heavily investigated all over the world.

Static network modeling

As we have already noted, a static network consists of entities, such as humans, cells, metabolites, or non-biological objects. These entities are generically called *nodes* or *vertices*, and connections between them are called *edges*. The adjective *static* signifies that the structure of the network does not change over time. Networks are often represented as *graphs* (Figure 10), and there is an entire field of mathematics and computer science, called *graph theory*, that addresses these structures. Static networks or graphs can be formulated very efficiently as matrices, and most analyses of graphs can be performed with methods of linear algebra, which offers an enormous repertoire of tools.

The edges of a graph may or may not have a direction, thus distinguishing directed and undirected graphs (Figures 10(a) and (b)). Over the years, many metrics have been developed characterizing the connectivity of networks, because this connectivity indicates relationships between nodes, and it also shows how quickly material or information can flow through the network. For instance, in the network of a food web, the directed edges indicate the various species of prey available to a predatory

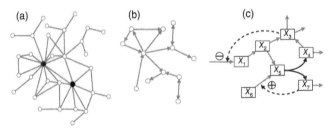

10. **Two typical networks and one regulated system: (a) undirected graph with two hubs (black dots), whose neighbors are densely connected; (b) directed graph, where some arrows are bidirectional; (c) system consisting of a network with regulation that is represented with a different type of arrows (dashed).**

species. A gene regulatory network exhibits which genes are up- or down-regulated together under different conditions, and an edge in this network offers information, or at least a clue, regarding proteins, called *transcription factors*, that control the expression of genes in different situations. Networks of these types are good examples of how models can complement the original data with novel information.

A particularly interesting type of network is one that contains many nodes but is connected in such a manner that it does not take many steps to move from any one node to any other node. A good example is the *small-world network* of airline routes, where relatively few nodes, called *hubs*, are heavily connected and most others are only connected to a few other nodes. In Europe, for instance, the major hubs include Amsterdam in the Netherlands, Charles de Gaulle in France, Frankfurt in Germany, and Heathrow in England. Each of them manages over 1,000 flights every day, and one can fly from one of these four airports almost everywhere within Europe nonstop or with maybe one stop. There are no direct connections between smaller airports, say, between Gothenburg in Sweden and Venice in Italy, but the hub structure makes it possible to reach the destination with only one stop in between (e.g. Charles de Gaulle or Amsterdam). If all nodes were connected more or less equally, and the network had the same total number of edges, a trip from Gothenburg to Venice would take many more stops. A small-world network with a similar structure of relatively few hubs with many edges describes metabolism, gene regulatory networks, food webs, and networks of neurons in the brain. An important metric quantifying the connectedness within a network is the *clustering coefficient* (CC), which measures how many connections actually exist among all neighbors of a node, divided by the maximally possible number of such connections. If CC is close to zero, the part of the network around this node is very sparse; if it is close to 1, then this node is a hub and many of the imaginable connections are actually there. Yet another metric is the *diameter* of a network, which looks at all shortest connections between any two nodes and identifies the longest (worst case; most steps) among these.

In contrast to static networks, the flow pattern of material in *regulated systems* changes over time. In order to allow for regulation, these systems contain arrows that come in two varieties: *transport* or *flux* arrows and *signal* arrows (Figure 10(c)). The *transport or movement of material* is depicted with heavy, solid arrows. Each of these arrows originates at a *source* box and points to a *target* or *sink* box. It is also possible that these arrows come from outside the system and therefore do not originate at a box or that they leave the system and therefore do not point to a box. For instance, animals may immigrate into a geographical area of interest (arrow from 'nowhere') or leave the area by emigration (arrow to somewhere else that is not of interest here). In further contrast to a typical static network graph (Figure 10(b)), an arrow may have two heads and/or two tails (Figure 10(c)). If the system represents a metabolic system, a double-headed arrow means that the metabolite at the *source* node (X_5) is split into two other molecules, represented by the *sink* nodes $(X_4$ and $X_7)$. A double-tailed arrow represents the case where two molecules are combined into a larger molecule. The second category of arrows represents *signals* that indicate the *flow of information*. A typical example is feedback inhibition. In Figure 10(c), several consecutive processes produce the variable X_3, and if there is enough X_3, then X_3 sends a signal to the production process of X_1 that says 'slow down' or 'stop.' Note that each such arrow originates at a box, but points to a transport arrow.

Dynamic modeling

The dynamic modeling branch of CSB uses a mathematical and computational toolset geared toward solving complicated puzzles associated with biological systems. The adjective *dynamic* is used when some aspect of the system changes with time. The general approach of dynamic modeling may be demonstrated best with a simple and idealized example, which nevertheless illustrates the general strategy within CSB.

Suppose we are interested in the growth of a bacterial culture. The culture starts with a small number of bacteria and while it grows, we measure regularly and report changes in population size. Such measurements can be accomplished with a variety of methods that do not disturb the bacteria. The result of our experiment is a series of numbers of bacteria that can be plotted against time (Figure 11(a)). In our idealized case, the plot suggests a clear trend showing that the population size increases during the experiment. In fact, closer inspection shows that the number of bacteria doubles about every twenty minutes. This regular increase makes sense because we know that bacteria multiply by division of one cell into two. So, we have 'hard data' (numbers of bacteria) and some additional information (doubling per cell division).

To continue our investigation as a dynamic modeling study, we now leave the realm of biology and jump into the world of mathematics. For this teleporting we evoke so-called *correspondence rules* that allow us to connect biological facts and findings with mathematical representations. The key player is the number of bacteria at time t, which we call $N(t)$. Mathematics tells us that the regular doubling of cells, from 1 to 2 to 4 to 8, and so on, can be represented with an exponential function of the type

$$E(t) = E_0 \, e^{r \cdot t}. \tag{1}$$

As indicated by the t in parentheses, $E(t)$ is a function of time. It contains one variable, t, and two unknown quantities, that are generically called *parameters*, namely, E_0 and r. E_0 is the value of the function at the beginning of the experiment, when $t = 0$, and we can measure this value. The parameter r determines how quickly the function increases and is therefore called the *growth rate*. Of course, we do not know the value of r, but there are

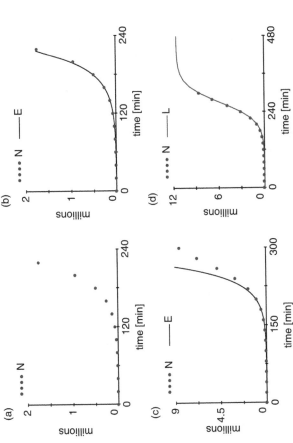

11. Results of investigating the growth of a bacterial population: (a) idealized measurements of the population size, $N(t)$, in units of millions. The bacteria divide every twenty minutes; (b) an exponential function (line) matches the observations (dots) very well; (c) continuing the measurements for a longer time period demonstrates that the exponential model (line) becomes quite inaccurate; (d) a logistic model (line) fits the same data much better and allows us to compute the carrying capacity of the system.

regression algorithms, with which computers determine the value for r so that the model results are closest to the measured data. In our case, such a *parameter estimation* algorithm would spit out $r = 0.03466$ and, if we could not actually measure it, it would also give us $E_0 = 1$ (in the units of the y-axis). For instance, 1 unit could equal 1 million bacteria. With these settings, the exponential function matches the data very well (Figure 11(b)). We call this function our *dynamic model*. This exponential growth model has been around since 1798, when it was proposed by the English clergyman Thomas Robert Malthus, who was also a scholar in economics and demography.

So far so good. Now suppose we would like to know whether the bacterial culture will continue to grow in the same manner or whether there might be a limit. To find out, we continue our measurements and again superimpose the data with the same exponential function model. The result shows very clearly increasing discrepancies between data and model (Figure 11(c)): the model is no longer appropriate.

In a situation like this, it is always advisable to consult the literature. Indeed, with a bit of sleuthing we find an old model for population growth that addresses this very issue. It is called the logistic growth model and was proposed in 1838 by the Belgian mathematician Pierre François Verhulst, who astutely asserted that any population growth must ultimately be limited by some resource. Indeed, even Malthus had already mentioned that population growth without limits could lead to famine and poverty, and these concerns are of course as valid today as they were 200 years ago. So, our observation is nothing new (and makes sense), although we do not know which resource is limiting growth and what the actual limit is. The logistic model for the size L of a population is formulated as a differential equation, namely

$$\frac{dL}{dt} = rL(1 - L/K).$$

(2)

There is no reason to be anxious about this type of equation; it just says that the rate of change of quantity L with time (the derivative on the left) is equal to some function (on the right). *Ordinary differential equations* (ODEs; lovingly called *O-D-E's* or *Diff-E-q's*) of this type tend to have a bad reputation, because they are more complicated than ordinary functions, such as the linear function $y = m\,x + b$ or the parabola $y = a\,x^2$. It is indeed true that we cannot directly plot what L in Eq. (2) looks like as a function of time. However, two facts should alleviate your uneasiness. First, there are uncounted computer programs that extract L as a function of time from the ODE (Eq. (2)); so, we do not have to worry about this aspect. Second, it is very often easier to describe how something changes, and what drives or modulates these changes, than explicitly to formulate this process as a function of time.

As a generic demonstration, suppose X is a component of some system, and we know that it is produced by some process and that it disappears because it is transported out of the system. It is difficult to describe the exact value of X at some future time point, but it is easy to formulate the change in X as an ODE. Namely, the change over time is mathematically defined as its derivative of X with respect to time (dX/dt) and we can therefore write

$$\frac{dX}{dt} = \text{Production} - \text{Transport-out} \qquad (3)$$

Voilà! We still have to formulate the production and transport processes in mathematical terms, but these can often be characterized from measurements, observations, or reasonable assumptions. Suppose the production is steady over time in the sense that so many units of X are produced in any given time period; say this constant rate of production has the value p. Furthermore, a transport process in biology is often proportional to X (the more there is, the more is transported). If so, Transport-out $= q\,X$, where q is some positive parameter that quantifies the speed of the transport process. Taken together, we have formulated the ODE

$$\frac{dX}{dt} = p - qX. \tag{4}$$

As soon as we have specified values for p, q, and the initial value of X at the beginning of the experiment (how many units of X are already there), a computer algorithm immediately produces X for every time point we want. In a simple case like Eq. (4), it does so much faster than the blink of an eye. Of course, most differential equations are more complicated, but it is rare that they cannot be *solved* by the standard computer algorithms of the field. The message is that there is no reason to worry: an ODE simply describes how various processes affect the change (dX/dt) in a variable X over time. This description is our dynamic model. It dictates what the computer algorithm is supposed to calculate. The algorithm itself is part of the machinery that is maintained and improved by experts in computing.

Let's return to the bacteria in Eq. (2). The literature tells us that the parameter r is again the growth rate, while the parameter K represents what a mathematician calls the *asymptote* of the function, the value that the logistic process approaches after some time. Applying a parameter estimation algorithm to the data in Figure 11(c) results in $r = 0.03466$ and $L = 1$ at time $t = 0$. Is that a coincidence? No, there are no coincidences in mathematics. The interpretation is that the bacterial population at first grows exponentially. The mathematical rationale is the following. If L is much, much smaller than K, which is the case when the population is still small, then the term L/K is almost zero, so that $(1 - L/K) \approx 1$. That means that Eq. (2) becomes

$$\frac{dL}{dt} \approx rL. \tag{5}$$

This ODE is actually a different-looking yet equivalent formulation of our original exponential function $E = E(t) = E_0\, e^{r \cdot t}$ in Eq. (1). To see this, compute the derivative of E with respect to t, which yields

$$\frac{dE}{dt} = r\,E_0 e^{r \cdot t}. \tag{6}$$

This result is already an ODE, because there is a derivative on the left-hand side and a function on the right-hand side. Now plugging in the definition of E on the right gives us $\frac{dE}{dt} = r\,E$. This result means that the exponential and the logistic models are essentially indistinguishable at the beginning of the experiment, when the bacterial population is very small; however, as the population grows, they differ more and more (Figures 11(b) and (c)). In fact, the exponential function grows without end toward infinity, whereas the growth of the bacteria in the logistic model approaches K, as predicted. The value of K is not obvious, but is easily computed from the ODE; its value in our example is about 11.552 (Figure 11(d)).

The results of the mathematical analysis are now transported back to the realm of biology, which interprets the asymptote K as the *carrying capacity*, which is the maximal population size the environment can sustain over time. The dynamic model demonstrates that we can predict how long and how far the population will grow, assuming that the logistic model formulation is appropriate.

While still within the realm of mathematics, we can also perform a number of *simulations*, a word derived from the Latin *similis*, meaning similar or analogous; simulations are discussed further in Chapter 6. In many ways, a simulation is like a video game that allows us to explore what happens if we change this or that in the dynamic model. A simulation uses a computer algorithm that takes a mathematical model of a biological system and allows us to manipulate its features. For instance, we could ask what happens if the rate constant r is changed from 0.03466 to, say, 0.05 or 5, or if L does not start at 1 but at 10. With a higher r, the population grows faster, but still approaches the same K. With a higher L, the population has a head start, but again grows toward K.

An interesting question is what happens if the population is greater in size than the carrying capacity K. That's an easy simulation demonstrating that the population decreases until it again reaches K. Teleporting back to biology: does that mathematical result make sense? Yes, it does. It could be that lots of animals immigrate or that the population faces habitat loss. If so, we expect the population to shrink. The logistic model quantifies this trend.

We might also be interested in knowing whether the temperature or the chemical composition of the growth medium affects the bacterial growth or the carrying capacity. Those are interesting questions, but they cannot be answered with the simple logistic model, because they clearly involve the physiology of the bacteria, which is not captured by this model. Such questions can be analyzed in CSB, but they would require a much more complicated model.

The bacterial growth example is greatly simplified, but it is a miniature version of reality that portrays the fundamental steps of CSB quite well. In actuality, many challenges lurk in the details. For instance, we may not really know all important features of the phenomenon, because nobody has investigated them yet or the data are incomplete or simply too inaccurate. In contrast to physics, where we have a lot of laws, such as the law of gravity or laws governing electromagnetism, biology has almost no known laws (see Chapter 7), which implies that we must always search for mathematical representations that we assume to be suitable. We saw already that the exponential model initially looked good but then failed us. What would we have done without Verhulst's logistic model? Well, we could have searched for other functions in the literature or even constructed our own, which of course requires some experience and background knowledge. Another challenge is to determine the best possible set of parameter values for our model. In the example, the numbers magically appeared, but in reality, the search for optimal values is often difficult. Again, many computer algorithms have been created for this purpose. Finally, it is not always obvious what analyses and simulations would be most informative and yield the most insights.

These challenges may seem to put a damper on CSB, but they actually create some of the excitement and appeal of the field. Unlike a computer game, nobody in the entire world may know the correct answer to a complicated research question. The mundane aspects of CSB are sometimes frustrating, but the research process overall can be a tantalizing journey of exploring the unknown as a true pioneer.

The new scientific method of CSB

The simple example of the bacterial population already demonstrated that CSB straddles two worlds: the actual, real world of bugs and flowers and diseases, and the abstract world of math and symbols and number crunching. This shuttling between worlds makes it clear that the research process in CSB is very different from the scientific method of the 17th century and also from the scientific method of the experimental world of —omics that we discussed in Chapter 3. It is summarized in Table 3.

As in the traditional scientific method, the typical investigation in CSB begins in the realm of biology or medicine, where the researcher formulates an interesting question or a specific hypothesis that might be worth the research effort. Of course, uncounted questions are possible, and it is very important to gauge whether enough data and accompanying information exist and are available to launch the modelling effort. If so, it is time to collect these data, along with contextual information, and organize all that's known and pertinent in a database that is dedicated to the project. An important decision to be made during this initial exploration is how much context needs to be included in the model.

Table 3. The scientific method of computational systems biology

Real world of biology	Shuttling between worlds by means of correspondence rules	Abstract world of mathematics
1. Identify interesting phenomenon to be analyzed		
2. Assess availability of sufficient information for a model analysis		
3. Collect data and associated information		
4. Determine context to be considered with the phenomenon		
	5. Translate phenomenon into an appropriate diagram with components and processes	
	6. Translate diagram into a mathematical model	
	7. Determine how biological information can be used in the model	
		8. Estimate parameter values
		9. Mathematically analyze model features as far as possible
		10. Implement model as computer code
		11. Diagnose internal consistency of the model and amend if needed
		12. Computationally analyze features of model structure
		13. Perform simulations
		14. Amend model if necessary and redo analyses
		15. Summarize results
	16. Translate mathematical results into biological insight	
17. Interpret results within the context of the phenomenon		
18. If results are not convincing, return to step 3		
19. If results are interesting, disseminate new information		
20. Generate new hypotheses; if these are interesting, return to Step 1		

So far, the steps of the scientific method of CSB are not all that different. But now we move over into the world of mathematics, by means of correspondence rules, as we discussed above. Depending on the data, the correspondence rules take us to the field of ML, static network analysis, or dynamic modeling. In fact, the three often form a pipeline that uses data and other biomedical information as input and adds new insights with every step until new understanding is created.

The correspondence rules are definitions of variables and formulations of processes, which are visualized with a diagram that shows the variables as boxes and the processes as arrows, either representing movement of material or the transmission of signals (see the earlier Figure 10(c)). All systems in biology contain such signals. It may seem like a trivial step to design such a diagram from biological information, but that impression is wrong. Indeed, the details of this conversion of real biology into boxes and arrows require much thought because the entire following analysis depends on this diagram being adequate.

Directly related to designing the diagram is the decision of how each and every process in the model should be formulated mathematically. Here, the researcher has two choices. One can either use default representations (like the exponential or logistic models above) that have been used many times before and hope that they are appropriate, or one may launch an investigation into which types of representations may be best for the given purpose. Obviously, the former option is easier, the latter more complicated, and there is no one-size-fits-all recipe.

Each equation contains parameters, such as the growth rate and the carrying capacity in the bacterial population case, and it is necessary to explore how these may be determined. If they are already published somewhere, life is easy. For most investigations, however, that is not the case, and we need to determine them from available data with computational methods.

This step of parameter estimation can be the most difficult and time-consuming part of the entire mathematical analysis.

With a good diagram and all equations and their parameter values specified, the first draft model is complete. While it may be possible to do some analyses of this model with paper and pencil, the next step is usually the implementation of the model in computer code that allows fast analyses. The first task at hand is the diagnosis of the model, where we ask: even if the model is presumably not 100 per cent correct, can we ensure that it is not obviously wrong? These diagnostics are executed with a mix of mathematical and computational analyses. In the former, we use methods of calculus, linear algebra, and other fields of mathematics, while the latter uses computer code that determines solutions with specialized algorithms. In a frequent scenario, even small numerical changes in a parameter value drastically change what the model does. This observation is reason for concern, because every biological system is constantly exposed to fluctuations in its internal milieu or external environment that change some of the parameter values a bit. If these fluctuations have huge effects, the model is probably wrong. There are several standard tests of diagnosis that the model must pass before we go on to analyses that tell us something new.

In many cases, most of the analyses are simulations, as we mentioned before and discuss in detail in Chapter 6. Many of these are explorations of *what-if scenarios*: What happens to the shape of the growth curve or to the carrying capacity if there is a large influx to the population? What happens to the population if the carrying capacity is changed? In a genetic model, we might explore what happens if a gene of interest is overexpressed several fold over its normal level. In a physiological model, we may ask what happens if blood pressure is ten points higher than normal. In the world of mathematics, these types of simulations usually go so fast that we can perform thousands of them. By contrast, if we ask 'is it possible that…' then we may have to analyze very many model settings, which can take a long time. There are no real guidelines for the number of simulations and other analyses to be performed, but a rule of thumb is the following: if reliable predictions can be made regarding untested scenarios most of the time, no more simulations need to be considered, because the correct predictions indicate a good understanding of the system. Otherwise, back to the computer!

As an alternative to targeted what-if simulations, so-called *Monte Carlo simulations* automatically screen thousands, if not millions, of different model settings and determine the overall repertoire of responses the model can produce. They also identify model settings leading to most likely, worst-case, or best-case scenarios, as well as unusual and unexpected responses.

All results from the mathematical and computational analyses are teleported back to the realm of biology, again via correspondence rules. Within this realm, they are interpreted, and questions may go back and forth between the realms of biology and math for a while. Ultimately, the analysis yields new biological insights, explanations, or predictions regarding so-far untested situations, which usually must be *validated* with targeted laboratory experiments that confirm they are correct. Mathematical analyses also often suggest new hypotheses regarding the specific roles of some of the system variables or processes. Some of these may be tested with computational analyses, while others require laboratory experiments. Confirmation is of course desired, but even if the validation experiments do not support the predictions, they can be very informative, as they may provide clues regarding unknowns in the system and can be converted into new processes or variables in the dynamic model.

At the end of the process, substantial new insights are disseminated among the research community and perhaps more widely. These insights open up new avenues of research and suggest new questions that initiate the next round of research.

Chapter 5

Interdependencies of biological systems

Even the simplest of organisms rely on uncounted molecular processes that allow them to thrive, propagate, and respond to threats from predators, parasites, and the environment. They possess thousands of genes defining each organism's repertoire of functions and features; proteins that provide structure, control biochemical reactions, and serve as means of transporting molecules; metabolites for energy and a myriad other purposes; as well as many, many signaling mechanisms that regulate all aspects of life. Pondering the immense range and diversity of these interacting processes, one might come to the sobering conclusion that any attempt to understand biological systems must be futile. But there is hope! While we may be dazzled by the variety of differences, there are also very strong similarities in the manner in which cells and organisms are organized. At the molecular level, much of this organization is captured in the *central dogma*.

Proposed first by Nobel Laureate Francis Crick half a century ago, the central dogma holds that the flow of information in cells starts with genes (DNA sequences), which are *transcribed* into messenger RNA (mRNA), which in turn is *translated* into proteins. Proteins play structural and signaling roles and control the conversion of food into energy and into metabolites, that is, the numerous and diverse chemical compounds required for life (Figure 12(a)). In a coarse sense, this simple version of the central dogma is still true today. However, over the past decades we have come to appreciate that the information flow is regulated in multiple ways. We now know that some proteins serve as *transcription factors*, which control the transcription of genes into mRNAs. Short strands of another type of RNA can regulate or even stop the transcription of specific DNA sequences. Metabolites can trigger or prevent the expression of specific genes. And we have learned that the translation of mRNAs into proteins and the activation of these proteins can be controlled in numerous ways and by a variety of molecules. The consequence of these insights is that the simple one-way process represented by the original central dogma is in truth much more complicated: information flows through a highly regulated system with many feedback signals (Figure 12(b)).

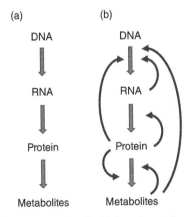

12. The central dogma of molecular biology then and now: (a) as originally proposed, the central dogma stated that information flows from DNA to RNA to proteins, some of which control the conversions among metabolites; (b) we now know that the information flow (straight arrows) is richly regulated by RNA, proteins, and metabolites through a variety of molecular mechanisms (curved arrows). Not shown is the process of reverse transcription, which in specific cases, such as some viruses, generates DNA from RNA.

Over the past fifty years, our knowledge of the molecular machinery of life has greatly increased, but when we explore the innumerable variations on each topic, we may be blinded by the diversity of molecular details among different species. However, we have also learned that the same fundamental principles of the central dogma hold in the vast majority of cases, from bacteria to plants, to humans, although there are exceptions here and there. This consistency in principles offers us the chance to gain some traction toward understanding how life works.

The central dogma dissects the complex machinery of life into distinct subsystems, which interact in ways that are often quite well understood, although nature always has surprises in store. The generic strategy for developing computational models of molecular systems is therefore, first, to capture the dynamics of the component subsystems and, second, to address interactions among them. The interactions between molecular systems are challenging to investigate, but they are very important and exciting, because they often lead to higher-level physiological systems and whole-organism systems. They also drive the responses of organisms to parasites, lead to mixed populations of different species, and govern even the largest ecological systems on Earth.

Gene systems

Genes are segments of DNA that contain most of the information needed to form an organism. Stretched out, the DNA of a single human cell would be about 2 meters long, but if all DNA molecules in the human body could be strung out, the total length would be about twice the width of our solar system. Mind boggling! The DNA in every human cell is curled up and wound around specific proteins, called *histones*, which makes it possible to fit the entire DNA into the nucleus of the cell. One of the amazing facts about DNA is that all cells of a higher animal, except for red blood cells, platelets, sperm and egg cells, have exactly the same DNA content. However, a skin cell is obviously very different from a brain cell. Also interesting, only about 2 per cent of our DNA consists of genes that ultimately code for proteins. The remaining 98 per cent were at some point arrogantly called 'junk DNA,' but we have learned that at least some of this DNA codes for small RNAs that play important roles in the regulation of gene transcription.

Great technological advances over the past decades have made it possible to *sequence* DNA very efficiently. In other words, we can determine the sequence of DNA's building blocks, A, C, G, T, for very large amounts of DNA and do it within an astonishingly short period of time. The letters are abbreviations for the four *nucleotides* adenine, cytosine, guanine, and thymine. DNA is *transcribed* into messenger RNA (mRNA), which is subsequently *translated* into protein. In these two processes of transcription and translation, combinations of three DNA nucleotides uniquely represent three RNA nucleotides, which in turn represent one of the amino acids that will ultimately make up this protein. As a consequence, the amino acid sequence of a protein can be read off the nucleotide sequence of the DNA. The very important *genetic code* is a complete list of assignments prescribing which combination of nucleotides corresponds to which amino acid. The use of only four building blocks seems quite limited, but it is sufficient to store information for an enormous number of different proteins (see next section, 'Proteins'). The computational comparison of gene sequences of different species is a major task in the field of bioinformatics.

Genes do not interact directly with each other, so they do not constitute networks or systems in a pure sense. But their indirect interactions are very interesting. For instance, it is important to know which groups of genes in multicellular organisms are transcribed under particular conditions and in which cells. This type of coordination is accomplished by transcription factors and a number of regulators, which are usually proteins or RNAs that themselves are the products of gene transcription. In other words, an arrow from A to B in a gene regulatory network diagram typically means that Gene A codes for the production of a transcription factor or regulator that affects the transcription of Gene B. In fact, transcription factors are frequently needed to express genes coding for other transcription factors, thereby creating hierarchical networks of transcription factors, and the same is true for networks of regulatory RNAs. We still do not fully understand these intricate systems.

As an example, Figure 13 displays a common plant response. The perception of drought stress leads to the activation of several pathways that are either dependent or independent of a signaling compound, called abscisic acid (ABA). These pathways activate various transcription factors (rounded rectangles), which interact with each other and ultimately lead to the expression of different sets of genes (hexagons). The arrows shown in the figure often involve several steps, such as the expression of a gene, its transcription and translation.

Understanding the functionality of these types of networks is an important task of CSB, but because they cannot be observed directly, they must be inferred from data with methods of ML (Chapter 4). The key step of a computational *network inference* is a sophisticated statistical analysis of data indicating that certain genes are *co-expressed*, which means they are transcribed into RNA under the same or similar conditions, such as heat, pressure, or exposure to some chemical in the case of a bacterial cell. If the same two genes are often co-expressed or silenced at the same time, the hypothesis is that they are controlled by the same transcription factor(s) and regulator(s). A validation of this statistically inferred regulation is often the discovery that the two genes contain the same short initial DNA sequence, to which the transcription factors and other regulators bind.

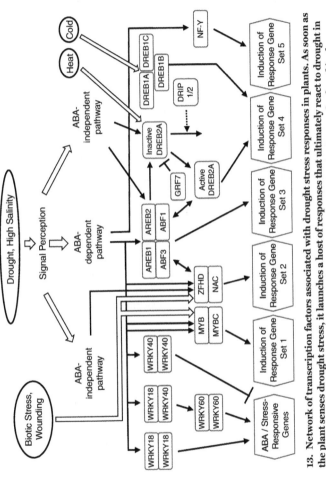

13. Network of transcription factors associated with drought stress responses in plants. As soon as the plant senses drought stress, it launches a host of responses that ultimately react to drought in an appropriate fashion. Details are not important for this illustration but can be found in the literature.

Proteins

Every organism is totally dependent on its genes because they contain most of the information for survival and propagation. But information alone is not sufficient, and actual work is required to thrive. Proteins are the true workhorses for most of the jobs an organism requires. They are at the core of most biophysical structures; they serve as transport engines, manage all the biochemistry that converts food into energy and into the compounds the body needs, and play numerous other roles (Figure 14).

Proteins are large molecules consisting of long chains of chemical building blocks, called *amino acids*. Nature uses twenty different amino acids for proteins, which does not sound like much, but permits a huge number of different proteins. Consider the potential number of combinations for proteins that are exactly 500 amino acids long! A full discussion of the variety of proteins and their roles is beyond the scope of this book, and we will instead focus closely on protein activities directly involved with the modeling of biological systems. Most current research activities in this restricted domain fall into three categories: availability and location, chemical structure, and function.

Research in the first category addresses questions regarding the amounts and localization of proteins under different conditions. The relatively new fields of proteomics and molecular imaging pursue these issues on a large scale. A typical question is: how do the amounts of proteins differ between a cancer cell and a normal cell of the same tissue? Another question may be: if the physical or chemical conditions in a cell's environment change, do some proteins move to different locations, such as the nucleus or the membrane? If so, which specific proteins move, where do they move, and what are the consequences? As an example, a chemical signal reaching the outside of a cell can alter the shape of a specific *receptor*, which activates an internal signaling process, which

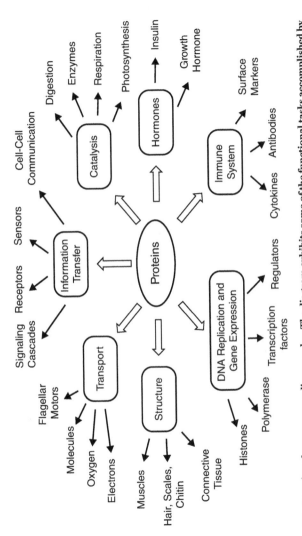

14. Proteins play numerous diverse roles. The diagram exhibits some of the functional tasks accomplished by proteins. The astonishing diversity of tasks is possible because of the almost unlimited number of configurations of the twenty amino acids that make up all proteins.

ultimately causes a protein to move from the cytosol into the nucleus, where it alters the expression of specific genes (see next section, 'Signaling'). This sequence of events happens in our bodies thousands of times every second. Computational models of such chains of causes and effects are important, as they shed light on the control structures within cells. They are either based on complicated statistics and ML that analyze large signaling datasets or with dynamic ODE models, which we discussed in Chapter 4.

The second category addresses questions of protein structure. Each amino acid in a protein has its particular chemical and physical properties, depending on its atomic composition and electrical charge, which may be positive, negative, or neutral. If many amino acids are linked together in a chain, these properties cause the protein to bend and fold into complicated three-dimensional structures, such as spirals and barrels (Figure 15). This three-dimensional structure, together with the surrounding chemical milieu, is responsible for the function of the protein. It would therefore be very helpful to predict a protein's

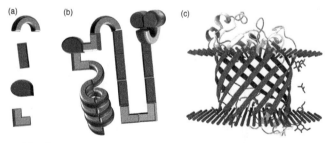

15. The shape of a protein is determined by the properties of its amino acids: (a) the four abstract shapes represent different amino acids, which can be linked to each other at their small flat faces; (b) the order of the amino acids, together with the surrounding chemical milieu, determines the three-dimensional shape of a protein, and thereby its function; (c) the *E. coli* enzyme phospholipase A has the shape of a barrel.

three-dimensional structure from its sequence of amino acids, which we typically know from the gene sequence that codes for it. But as easy as this may sound, even the largest computers have great difficulties with this type of *structure prediction*, which therefore remains a 'grand challenge' in systems biology. Related to this task is the prediction of how a small molecule, such as a drug, will bind to a protein, how the protein structure will change as a result, and how this change will alter the activity of a protein. Clearly, this line of research is of great interest to academia and the pharmaceutical industry.

The third category of protein research focuses on the function—or malfunction—of proteins. For instance, proteins are critical players in essentially all diseases, whether as signal receptors, enzymes, transporters, or something else. Computational models therefore often focus on the consequences of protein alterations rather than on the proteins themselves. We will consider some of these functions in the following sections on signaling and metabolic pathways.

While our focus here is on the central dogma, we must not forget that proteins are also responsible for the transport of compounds between cells and through the bloodstream, that they drive our immune system, and that they permit the function of muscles, thereby allowing movement, including the regular contractions and relaxations that govern our heart, lungs, and digestive system.

Signaling

The fine-tuned coordination of systems within each cell or organism requires that their parts must 'know' what is happening in other parts, at least to some degree. This sharing of information is called *signaling*, *signal transduction*, or *signal transmission*. Generically, a signal is sent by a source and received and correctly interpreted by the intended target. Touching a hot stove is perceived by sensors in the fingers, the information is sent via nerves to the spinal cord, which sends back signals to the arm muscles instructing them to remove the hand from the stove; all in a split second. This type of signaling through nerves involves chemical and electrical processes. At the cellular level, signaling can be achieved with proteins, lipids, small molecules like calcium, or electricity. Usually, signal transduction involves an entire system of components and processes.

A widely found signal transduction system is the combination of a receptor with a protein-based *signaling cascade*. The receptor itself is located in the cell membrane with one part of it sticking outside the cell and the opposite part reaching into the inside. The outside component is able to bind to appropriate signaling molecules outside of the cell, which are collectively called *ligands*. These ligands may be of very different types but are specific for a given signaling task. Once a ligand binds, the inside *conformation* (molecular shape) of the receptor changes, thereby serving as an internal signal turning on a signaling cascade.

The best-known example is the so-called MAP-Kinase cascade, in which phosphate groups are attached to proteins, thereby activating them (Figure 16). It consists of three layers. The activity at one level of the cascade triggers the next level, and the final product facilitates some action, such as the movement of a transcription factor into the nucleus, where it causes the expression of target genes. Many such cascades work side by side inside the cell and *cross-talk*, which means that one cascade may activate another cascade. This cross-talk permits the appropriate interpretation of very many signaling scenarios.

Computational research in this area falls into two groups. In the first group, the goal is to determine how the various cascades are composed, how they interact, and how signals travel throughout a cell. These studies usually involve complex statistical ML.

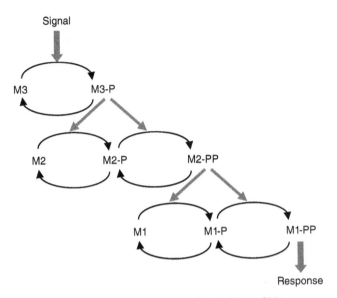

16. Typical protein-based signaling cascade. M1, M2, and M3 are proteins that are activated when one or two phosphate groups (P) attach to them. The entire cascade becomes active if a specific signal of sufficient strength is present. The ultimate outcome is the triggering of a specific response, such as the movement of a transcription factor into the nucleus, where it causes the expression of one or more genes.

The second group attempts to understand the dynamics of known cascades with ODE models.

Metabolic pathways

Metabolic pathways are chains of biochemical reactions that convert metabolites into other metabolites. As an example, a chain of ten reactions converts glucose from food into the common intermediate metabolite pyruvate, which is subsequently used in numerous other reactions. Each biochemical reaction uses a *substrate* (e.g. glucose) and converts it into a *product* (e.g. pyruvate), which then becomes the substrate of the next reaction (Figure 17). As a frequent variant, a reaction may have two substrates and/or generate two products. Along the way, new necessary metabolites are generated, energy is extracted from the metabolites, some metabolites are stored for times of need, and low-energy, unwanted, or toxic compounds are excreted through the liver and kidneys. Specialized *enzymes* facilitate most of these reactions by acting as catalysts.

An example is the pentose-phosphate pathway, which branches off glycolysis (Figure 17). The common initial substrate, glucose, is converted into G6P. For glycolysis, G6P is further converted into fructose 6-phosphate, but some of the same G6P enters the pentose-phosphate pathway toward ribulose 5-phosphate. How much G6P moves either way depends on numerous details of the

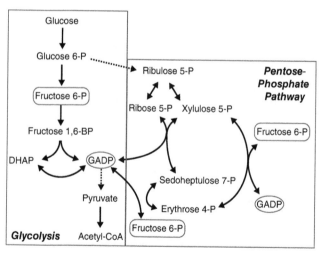

17. **Simplified glycolysis and pentose-phosphate pathway. The universal occurrence of glycolysis suggests that this paradigm of a metabolic pathway is ancient. Here, it is shown to interact with the similarly crucial pentose-phosphate pathway. Dotted arrows represent chains of reactions. Framed compounds are involved in two or more locations.**

biochemical and physiological status of a cell. As a consequence, the functionality even of this simple pathway is not easy to predict. And there are other factors complicating such predictions: for instance, fructose 6-phosphate and GADP (glyceraldehyde 3-phosphate) participate in other reactions, and the regulation of the pathway is not even shown here.

Metabolites may be seen as the last components of the central dogma (Figure 12); indeed, the pathways converting them into each other are often the ultimate targets of changes in gene expression. Genes code for the fundamental information governing our bodies, RNAs are intermediaries, and proteins are the machines responsible for the smooth running of physiology, including metabolism. It is often at the level of metabolites that differences between health and disease manifest. Diabetes is associated with an imbalance of glucose in the blood stream. Parkinson's disease results from low levels of the brain chemical dopamine, and cancer cells rewire their metabolism for energy production and proliferation. Metabolites often go unnoticed if everything is well, but they come to the fore if we are afflicted by disease.

Every organism contains hundreds, if not thousands of different metabolites; some are very common, others are rare. Plants are masters in producing rare metabolites, which contribute to their colors, flavors, and defenses against animals that may want to eat them. Microbes produce many of the same metabolites we humans use, but some bacteria can produce antibiotics like erythromycin, and penicillin is produced by a tiny fungus.

While plants can use sunlight to generate carbohydrates, animals must produce all necessary metabolites by chemically converting compounds from food. This conversion can be rather simple, quite complicated, or impossible. For instance, our bodies easily generate fats from carbohydrates, but they cannot produce vitamins, so we have to eat plant materials (or, nowadays, dietary supplements) that contain them.

Most biochemical reactions are executed by enzymes, whose amounts and chemical and physical features determine how fast a substrate is converted into a product. Most enzymes are very specific for the task and execute only a single reaction or at most a few reactions with similar substrates, although there are exceptions. Figure 17 shows the example of glycolysis and a related pathway that generates sugars with five carbon atoms, called *pentoses*. Large databases like KEGG (Kyoto Encyclopedia of Genes and Genomes) and MetaCyc contain electronic catalogues of most of the known metabolic pathways, along with their enzymes. As an example, Figure 18 shows parts of the detailed KEGG representation of the right half of Figure 17.

From the viewpoint of controlling a system, like a metabolic pathway system, a very important feature is the option of regulating the *fluxes*, that is, the amounts of material flowing through each reaction at a given time. A prominent tool is feedback inhibition: If a lot of end product of a pathway is already available, the end product itself can send a signal to the first step of the pathway, 'instructing' it to slow down or even stop (Figure 19). This instruction is accomplished by affecting the enzyme that is responsible for the first step of the pathway. Two very frequent mechanisms are *competitive inhibition* and *allosteric inhibition*. In the former case, the end product has a chemical structure that is so similar to the substrate of the first step, that it competes with it for the enzyme, thereby tying up some of the available enzyme molecules and slowing down the reaction. In the latter case, the end product, or some other metabolite, can bind to some location on the surface of the enzyme and by doing so change the activity of the enzyme with respect to the substrate. In both cases, the enzyme activity is reduced or stopped.

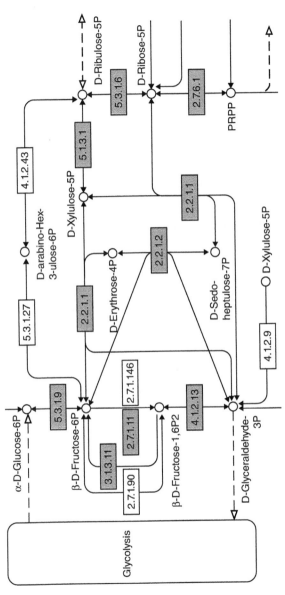

18. Pentose-phosphate pathway, as represented in the database KEGG. Some of the right part of Figure 17 is shown again in more realistic detail, even though regulation is still not included. The numbers in boxes are internationally agreed identifiers for enzymes.

71

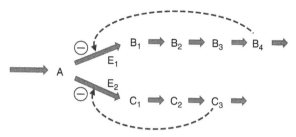

19. Branched metabolic pathway with regulation. Substrate A can be used for the production of B_1, B_2,..., as well as C_1, C_2,.... The two branches are separately controlled by the enzymes E_1 and E_2, and inhibiting (or activating) these allows the cell to channel more of A into one pathway or the other.

Feedback inhibition is particularly useful at a branch in the pathway, where a metabolite can serve as the substrate for two or more pathways (see Chapter 2 and Figure 19). Each branch is under the control of a different enzyme, and if the activity of the enzyme of one branch is reduced, more material will flow into the other branch. This rechanneling is vital for the cell if, all of a sudden, one of the end products is in high demand. Increases in enzyme activity can be achieved quickly through other metabolites that specifically interact with the enzyme.

As an indication of how important regulation is, other control systems are available to a cell. If a more permanent reduction is needed, the enzyme can be deactivated by proteins uniquely tasked with this process. Often, the deactivation is accomplished by adding or removing a small molecule like a phosphate group. Moreover, the amount of enzyme is governed by the balance between production and destruction. One control option therefore is to increase the amount of available enzyme, and because almost all enzymes are proteins, the corresponding gene expression needs to be cranked up, which leads to increased mRNA, ultimately resulting in more enzyme. The cell may also remove enzyme molecules with the help of proteins, called *proteases*, that are specifically tailored for this purpose.

Computational models of metabolic pathways generically consist of: *pools*, which can be imagined as containers for molecules; *fluxes*, which are quantities of material flowing among the pools through the reactions; and *signals*, which regulate some of the fluxes. Metabolic models commonly address two features of the actual pathway. The first is the *steady state*, where all metabolite concentrations remain the same, even though material is flowing through the system. Many healthy pathways operate close to this state of *homeostasis*. A great modeling advantage of focusing on the steady state is that many associated features can be obtained with numerous, powerful methods of linear algebra and optimization. For instance, it is possible to characterize the distribution of metabolic fluxes throughout the pathway system and to predict how these fluxes would change if one altered the activity of one or more enzymes. To some degree, it has become possible to reconstruct the entire metabolic system of an organism. These methods are becoming more and more popular in metabolic engineering, where an important goal is the use of microbes for the production of bulk materials like industrial alcohol or of expensive compounds like insulin and prescription drugs (see Chapter 6). A beautiful example of combining metabolic engineering with systems and synthetic biology is the production of the antimalarial drug artemisinin in a yeast strain that had been specifically altered and augmented for this purpose.

The second feature of a metabolic pathway system is its dynamics; the overriding question is: how do the metabolite concentrations of this pathway change over time? In this case, linear algebra is insufficient and the approach of choice is the use of ODEs. Help here comes from the fact that metabolic pathways permit strict bookkeeping: The exact amount of material that leaves a metabolite pool must enter one or more of the other pools; no material gets lost or magically appears. This constraint is very beneficial for modeling, as it restricts the choice of model settings.

Interactions among molecular systems

Under normal conditions, hundreds of biochemical reactions perpetually generate the same metabolic products the body needs. However, when the organism experiences stress, it must respond to prevent damage. This response requires changes in the concentration profile of specific metabolites that are protective or otherwise beneficial under adverse conditions. Let us consider a specific example as an illustration.

Baker's yeast, which we use for baking bread and making beer, likes a temperature of about 25°C. When the temperature in the environment increases to, say, 35°C, it causes *heat stress*, to which the yeast cell must adapt (Figure 20). This adaptation involves

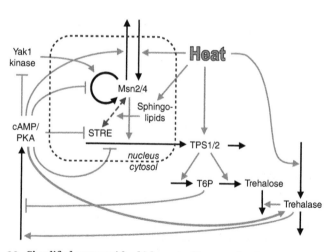

20. Simplified system with which yeast cells respond to heat stress. The rounded square represents the nucleus; black arrows indicate altered gene expression or metabolic reactions, while grey arrows show activating effects and grey 'T-lines' represent inhibiting effects. The dashed arrow signifies the interaction between the transcription factors MSN2/4 and the stress element STRE.

several metabolic pathways and leads to the production of specific heat stress response metabolites and proteins, including the transcription factors MSN2/4 that interact with the stress element STRE, a short segment of DNA preceding the genes TPS1/2 and allowing transcription factors like MSN2/4 to initiate gene expression. As we already mentioned in Chapter 2, a hallmark metabolite is the sugar trehalose, which the cell produces in enormous amounts to protect important cellular structures, but only under stress conditions. Another hallmark is the production of sphingolipids, which are structural components of membranes but also have signaling functions that can trigger the expression of response genes.

Altering the amounts of specific metabolites requires well-orchestrated adaptation strategies. First, the cell needs a sensor system that detects the temperature increase and transmits this information to molecular 'response teams.' In the case of yeast at 35°C, one sensing system consists of three enzymes producing and degrading trehalose (Figure 5). Heat increases the activities of the producing enzymes and reduces the activity of the degrading enzyme. The result is the production of large quantities of trehalose. A second sensing and signaling system is the generation of a specific sphingolipid that effects changes in the expression of genes that code for specific response proteins. In addition to these systems, heat-shock proteins are activated, and much of the physiology of the yeast cell changes in response.

To summarize, responses to substantially altered situations are complicated. They are achieved through the fine-tuned coordination of different molecular or physiological systems operating in parallel. They require sensors and information transmission processes that trigger mechanisms connecting genes, proteins, and metabolites with the overall physiology of the organism and equipping them to respond effectively to the numerous stresses a cell might experience in nature.

Systems spanning many scales

The heat stress example indicates that physiological processes require the coordination of many molecular systems. The situation becomes even more challenging if we consider macroscopic or global processes that are driven by molecular events. Consider the following all-important example.

Astronauts and cameras on unmanned spaceships have taken photos that show our planet Earth from the distance. Most of Earth looks beautifully blue, due to the vast oceans that cover over 70 per cent of its surface and provide over 99 per cent of the total space for life. Contrast this vast expanse to one of its smallest inhabitants, the cyanobacterium *Prochlorococcus*, which measures less than a micrometer in diameter; to put this size in perspective, the naked human eye can recognize objects down to a size of about 60 micrometers. A single milliliter of surface seawater can contain more than 100,000 of these fellows, which adds up to an estimated world population of 10^{27} *Prochlorococci*. This huge population of about one octillion organisms is believed to produce 20 per cent of all oxygen on Earth. In addition to allowing us to breathe, *Prochlorococcus* and other types of phytoplankton (literally 'drifting plants') are food sources for very small animals, the zooplankton ('drifting animals'), which consists of tiny crustaceans and a multitude of larvae of larger animals like crabs, squid, and fish. This small zooplankton is food for larger, predatory zooplankton, like krill, clams, and sponges, which are eaten by many fishes, large and small, and even by the largest animals on earth, such as the huge blue whales that strain floating biomass through specialized filtering structures. Of course, small fish, in turn, are food for marine mammals and larger fish, some of which we consume ourselves. According to the 2010 Census of Marine Life, the oceans are home to about 250,000 species of animals and plants, including over 15,000 fish species, and all these species interact with many others in often unknown ways.

Add to these numbers the enormous quantities of interacting marine microbes and viruses, which live in complex mixed-species communities and actually make up almost 90 per cent of the oceans' biomass, and there is no doubt that the food webs in our oceans are incredibly complex.

For the systems biologist, this situation poses a huge challenge. Clearly, we cannot study the entire ocean at once and with a resolution that accounts for every cyanobacterium. So, what is the solution? The honest answer is that we do not really know (yet). *Multi-scale modeling*, as it is called, is a research frontier that direly needs innovative, out-of-the-box ideas. One rather obvious approach that is widely being explored is a *modularization* of the large system. The *modules* are systems of a size and complexity that appear to be manageable, and interactions between modules are studied once the modules are more or less understood. Two critical issues with this approach are daunting. First, even minute perturbations in a single cyanobacterium, multiplied by one octillion, might have huge global effects. And second, some modules, at any scale, are likely to affect other modules at the same, higher, or lower levels, possibly leading to responses and adaptations throughout the global system. How can we effectively capture this connectedness and interdependence in computational models, if it vastly overwhelms our minds? Can we trust computers to put the pieces together correctly?

If your head is starting to spin, that is not surprising; in fact, it is natural. There are just too many wheels turning to keep track of what is going on. We often understand interactions between two items or among a small number. But we cannot scale up to systems of dozens of interacting components. Overwhelmingly complicated examples do not even have to be as large as the world's oceans. The immune system and brain are extremely puzzling, and processes like embryonic development and cancer metastasis are becoming a little bit clearer but will require much future investigation. CSB attempts to shed light on the details of these well-coordinated systems, so that we may eventually understand them and, if need be, manipulate them in a desired manner, for instance, from disease toward health.

Chapter 6
Simulators

Computational models can serve many purposes. We can use them to interpret data and to fill gaps in knowledge where we do not have data. We can extrapolate models to explore so-far untested scenarios or even situations that cannot be analyzed with experimental means, because we do not have the technical know-how or because experiments would be unethical, dangerous, or simply too time-consuming or expensive. We can test hypotheses about a system by setting up a model that focuses exclusively on the essential features of the system while omitting distracting details. In some cases, we can use these types of analyses to explain objectively why a certain system design that we observe in nature is actually superior to alternatives that at first seem equally good or even better (see Chapter 7). Sometimes models can even explain system responses that are puzzling or counterintuitive, as the story on glucose uptake by the bacterium *Lactococcus* in Chapter 2 demonstrated.

A particularly powerful application of a model is its use as a *system simulator*. We have all heard of flight simulators (Figure 21). They look exactly like actual airplane cockpits and allow a pilot-in-training to practice routine tasks such as taking off and landing, but they also realistically mimic situations with which the pilot will hopefully never have to cope in real life, such as the loss of one of the plane's jets or failure of its landing gear.

21. Flight simulator replicating the F-111 cockpit. The General Dynamics F-111 Aardvark is a US Air Force long-range strategic bomber, reconnaissance, and tactical strike aircraft.

Flight simulators have become indispensable teaching tools in about every pilot's training. They seem so real and are so effective because physicists and engineers understand the forces, factors, and processes influencing a plane very well. For instance, they have developed reliable computer models that accurately simulate air-flow around the fuselage and across the wings and take into account all forces affecting the plane as well as forces generated by the plane. These detailed models are modules within a complicated 'supermodel' that permits the artificial creation of very realistic-looking scenarios to which the learning pilot must respond. Each response by the pilot is fed into the appropriate modules, and the supermodel computes and visualizes how the plane reacts. All this happens so fast that the actions of the pilot and the plane's responses appear to be real.

An emerging branch of CSB strives to develop simulators of this genre for complex systems in biology and medicine. The premier example is a disease simulator. In direct analogy to a flight simulator, it is easy to imagine how the medical student of the future will learn the outcomes of various treatments of a disease and the typical, but also very rare responses from a small number of patients. The brain behind the simulator will be a complicated computer algorithm that is capable of creating realistic disease situations, and the medical student will be asked to assess these situations and order tests. The algorithm simulates these tests, returns the results to the student and then asks what to do next. The student's strategy is evaluated first and foremost in terms of health outcome, but also with respect to potential side effects and overlooked differential diagnoses. The evaluation will also cover unnecessarily wasted resources, alternative treatment plans that might have yielded the same curative results, possibly with fewer side effects, and estimate whether such alternatives might have been faster or cheaper. The same type of disease simulator is also likely to become the diagnostic tool of choice. Suppose a future patient has some health problem or a regular check-up. Blood, urine, and maybe other biological samples are taken and immediately analyzed with fast versions of traditional tests and with modern —omics and machine-learning tools. The data are fed directly into a comprehensive computer model, and this model not only identifies possible health problems, but also suggests a treatment, even though the doctor remains in charge and carries the responsibility for subsequent actions.

All this may sound wildly futuristic, but the first steps have been made. Why are we not there yet? If we can simulate the movement of a modern airplane and all the environmental factors influencing it, why can't we model diseases right now? The reason is again, in one word, complexity. Yes, an airplane is certainly complicated: one of the workhorses of the airline industry, Boeing's 737, consists of 367,000 parts; its 787 Dreamliner has about 2.3 million parts. Alas, the human body contains almost forty trillion cells, each with its own molecular inventory, and there are still very many molecules and processes we do not really understand well. And if we do not understand them, we cannot convert them into effective computer models. So, we are still far from designing realistic disease simulators with sufficient accuracy and reliability, especially if they are supposed to be personalized, but they are nonetheless an intriguing goal of CSB. In fact, the University of Illinois recently created a futuristic center dedicated to healthcare simulation.

With a bit of imagination, one can envision numerous other biological simulators, some of which are already under investigation, while others have not even come to anyone's mind. This chapter discusses nascent efforts toward such simulators. Before we proceed, a cautionary note is in order. One must always keep in mind that any results from simulations, even under the best conditions, are only as good as the assumptions made, and these are not always based on solid facts. They may even be biased by some scientific or political agenda, especially within the realms of health and of global environmental questions. Nonetheless, if current trends are any indication, simulations will be central tools for systems biology in the future.

Health and disease

What exactly is the difference between a healthy person and a sick person? That probably sounds like a silly question, but it really isn't. Maybe the sick person has a fever, but that is not always the case; think about migraines or diabetes. Maybe diseases come with high blood pressure. But is a person with blood pressure of 145/95 really sick, just because two numbers are a bit higher than the 'ideal' of 120/80? Should a person who lost a toe in an accident always be considered sick?

The truth of the matter is that it is very difficult to define, in a general, concise manner, what the concept of disease means, without using the concept of health, or vice versa. Complicating the formulation of crisp definitions is the obvious fact that there is a gradual transition from being and feeling healthy to feeling so-so, to feeling lousy, to actually being demonstrably sick. Where exactly is the threshold? Also, many processes in the body naturally slow down during aging, which further complicates our definitions of health and disease, as only young people might call old age a disease.

Thinking like a computational systems biologist may help us with an objective assessment of health and disease, at least in the form of a thought experiment. Not surprisingly, the CSB approach prescribes that we view the human body as a huge, complicated system. All processes in this system, whether they are physiological, metabolic, or associated with genes, neurons, or other components of the body, are represented with mathematical functions, many of which are incorporated in differential equations. We may not know most of these functions in detail, but that is not important for our thought experiment, although an actual disease simulator would need to know all those details. As we discussed in Chapter 4, every mathematical function in CSB contains variables, such as concentrations of metabolites or numbers of cells, as well as parameters that quantify the speed or magnitude of a process under investigation or generically capture the importance of each variable for the functioning of the system.

With this mindset, it is not hard to imagine that health may be characterized by parameter values that are all 'normal'; this state or condition of the body with normal parameter values is called *homeostasis*. By contrast, disease may be imagined as the consequence of one or more 'faulty' parameters in the system. Consider, as an example, that every heart beat depends on calcium flowing between two separate compartments within every heart muscle cell. If this flow decreases, the cell becomes less efficient, the contractions of the heart become weaker, and the heart as a whole does not fill with blood and empty as quickly as a young, healthy heart does. The consequences can be shortness of breath, the accumulation of fluids in some tissues, and possibly congestive heart failure. In our model of the system, the magnitude of calcium flow is represented by a parameter, and dialing down this parameter seamlessly leads from health to disease. In fact, this reduction in calcium flow actually happens quite often in older people. So, aging can indeed look like a disease.

Very many processes operate in our bodies, day in, day out, and each process contains one or more parameters that could possibly derail. Looking at health in this manner, it is indeed a wonder that most people are basically healthy! Now, it is rather unlikely that something terrible happens if the value of a parameter is off by merely 1 per cent; nature is too tolerant and resilient for such pettiness. But what about 5 per cent, 50 per cent, 500 per cent? As in so many other cases in biology and medicine, the answers depend on the specifics of the context. Some parameters are very sensitive, so that even moderate changes can have far-reaching consequences, while others may vary a lot without causing many problems. In addition, it often happens that deviations in some parameters are compensated for by other parameters, which is typically the result of biological control circuits that are designed to keep the body running smoothly.

If parameters or variables are particularly sensitive, so that relatively small alterations are associated with a disease, they are called *biomarkers* for that disease. Expressed casually, biomarkers are parameters or variables in the human body that are frequently (although not always) altered if an individual is diseased. The great interest in biomarkers derives from the fact that alterations in biomarkers are often detectable *before* a disease manifests, so that they can be used as tools for predicting, with some reliability, whether a particular disease may be imminent, thereby allowing early interventions. In reality though, we seldom know enough biomarkers, and predictions only indicate whether the *probability of disease* is increased or even decreased. A good example is a mutation in one of the two so-called *BRCA* genes. Such a mutation is associated with an increased risk of breast and ovarian cancer but does not mean that every woman with this mutation will actually develop cancer. Also, many women without such a mutation develop cancer. So there is no one-to-one relationship.

Because nature cannot ensure that a parameter always has its optimal value, all parameters are allowed some slack. In other words, every parameter in the body has a normal range, within which its value does not really matter. Sensitive parameters have small ranges, whereas insensitive parameters have much larger ranges. Taken together, the normal ranges of all parameters can have very different widths, which we usually don't even know. If we focus on three biomarkers, their ranges collectively result in a rectangular 'normal' box (Figure 22(a)). Experience shows that combinations of extreme biomarker values are often not so good, and if we account for this observation, the corners of the box are no longer healthy and therefore should be avoided. If we cut them off our rectangular box, the result is something like a cut gemstone, which mathematicians call a *simplex* (Figure 22(b)). We cannot visualize a simplex in thousands of dimensions, which would reflect our *health domain*, but even the three-dimensional simplex provides us with a concept of health (inside) and disease (outside), and one can easily imagine that the

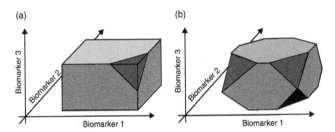

(a)   (b)

Biomarker 3
Biomarker 2
Biomarker 1

Biomarker 3
Biomarker 2
Biomarker 1

22. Visualization of health and disease as simplexes: (a) health is defined by the normal ranges of three biomarkers, that is, health states are located inside the three-dimensional grey box. If combinations of high values for Biomarkers 1 and 3 and a low value of Biomarker 2 are considered unhealthy, the corresponding corner (dark grey) is cut off; (b) if all combinations of extreme values are excluded, the result, in mathematical lingo, is a simplex.

further away a person is from the health simplex, the sicker he or she is likely to be.

Traditional medicine, of course, does not resort to simplexes, but usually focuses on the strongest driver of a disease. In the language of systems biology, this driver corresponds to the most influential biomarker outside its normal range. For instance, no matter what all the other parameter values are doing, a cholesterol level of about 200 milligrams per deciliter (mg/dL) is considered desirable for an adult, whereas a level of more than 240 mg/dL is deemed too high.

If we assert that a biomarker outside its normal range is a significant risk factor, we ignore the important fact that the body can compensate for many abnormalities. This compensation is usually accomplished with a complex system of regulatory mechanisms that counteract the consequences of rogue parameters. These mechanisms often alter other parameters in the system, with the result of more or less normal overall operation. In other words, not every parameter is truly important for health and disease, but some key parameters are, and these are particularly useful as biomarkers. If parameters leave their normal ranges, other parameters may be adjusted in an attempt to keep the system operational. Often these adjustments are sufficient, and we don't even notice that anything is wrong, but if they are not, we are faced with disease.

As an illustration, Figure 23 shows a simplified diagram for the production of purines, which are molecules needed to synthesize DNA, RNA, ATP, and other important compounds. Under healthy conditions in humans, the variables of the system have steady-state values of about $X_1 = 5$, $X_2 = 100$, $X_3 = 400$, $X_4 = 10$, $X_5 = 5$, and $X_6 = 100$. Quite a few diseases are associated with the malfunctioning of this pathway system. A particularly devastating example is Lesh-Nyhan syndrome, a severe neurological and behavioral abnormality, characterized by involuntary muscular

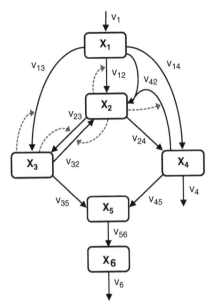

23. **Simplified purine pathway.** X's in boxes are variables representing biochemical compounds, v's represent enzymatic conversions, and the dotted arrows indicate inhibition signals. In Lesh-Nyhan syndrome, the activity of the enzyme that is associated with v_{31} and v_{42} is reduced to a low value like 10 per cent. The consequence is a severe perturbation in many of the metabolite concentrations of the pathway.

movements, mental retardation, and compulsive self-mutilation. The disease is caused by a strong reduction in the magnitudes of the enzymatic processes v_{31} and v_{42}. If these are operating at only 10 per cent of the normal capacity, the steady state of the pathway changes to about $X_1 = 9.5$, $X_2 = 98$, $X_3 = 233$, $X_4 = 18$, $X_5 = 10$, and $X_6 = 143$. So, in comparison to a healthy pathway, X_1, X_4, and X_5 are roughly doubled, X_6 is over 40 per cent too high, and X_3 is much too low. Interestingly, if we could adjust the magnitudes of other processes within the pathway, for instance $5 \times v_{12}$, $2 \times v_{23}$, $1.5 \times v_{45}$, $1.5 \times v_{56}$, $1.5 \times v_6$, the healthy steady state would essentially be restored. Unfortunately, the body cannot always implement such adjustments, but model analyses of such compensatory changes demonstrate how one might eventually approach the search for new treatments.

Some combinations of altered parameter values cannot be fully compensated, but nature tries hard to counteract detrimental perturbations. The result of this attempt is the frequent and puzzling observation that the 'same' disease, such as breast cancer, manifests quite differently in different individuals and that some drugs work for one person but not another. In the purine case, a moderately reduced activity of v_{31} and v_{42} can normally be compensated by the body, but if one person's metabolism cannot alter v_{23} for some reason, or if it has a secondary enzyme defect somewhere in the pathway, the disease looks very different in the two individuals.

Because we do not know the ranges of all parameters and the actual locations of their values within these ranges, differences in the manifestation of a disease are difficult to understand. Ideally, we would like to check all pertinent parameters simultaneously, but that is simply not possible at present. Nonetheless, the future has begun. As an example, a relatively straightforward gene expression test can predict whether particular treatments for childhood leukemia will be effective or not. These types of prognostic tests will become much more commonplace for severe diseases and offer hope for customized treatments. Also, several for-profit companies have begun to assemble systems models of specific diseases.

More generally, a modern goal of medicine is to redirect traditional treatment strategies toward *personalized medicine*, which will eventually replace the current one-size-fits-all approach with treatments or interventions that are custom-tailored for a specific patient. Systems biology is on its way to play a major role in this trend because the full complexities of health and disease simply cannot be captured without computational approaches.

This involvement of CSB is still in its infancy, but we can imagine how it might develop. The two main ingredients will be a detailed supermodel of the entire human body and a list of the normal ranges of every important parameter. Entering average values from these normal ranges into the supermodel provides an impression of a healthy 'average' person. Taking different combinations of values from within the health simplexes we discussed before will give us variations of health, while combinations with some values outside these ranges will reveal different disease patterns. Replacing healthy parameter values with personal values, measured in the same patient, will tell us how much the patient diverges from the normal state of health and suggest which countermeasures might be effective in restoring health.

Microbial factories

Microbes are everywhere. They live in soil and streams and on rotting food, and they even inhabit environments that seem very unlikely to permit life, such as ocean vents spewing out toxic gases and liquids that are so hot one would think all life would immediately come to a boil. Some deserts are extremely hot and dry, and it is hard to imagine any creature surviving for long. Yet, all imaginable niches are filled with bacteria, fungi, and very simple single-cell archaea, all adapted to whatever nature dictates and demands. Some 'extremophiles' actually thrive at 110–120°C, without being cooked; others live at about –18°C, without being frozen stiff. In contrast to our bodies, which consist of 70–80 per cent water, some desert bacteria can survive in suspended animation with just 1–2 per cent water content, waiting to bounce back as soon as water becomes available. All of us are host to over a thousand species of microbes in our guts, on our skins, and in our oral cavities, for an estimated total of about a hundred trillion, which is three times the number of our own cells. Considered that way, our bodies are really containers of microbes. The bacterial communities are essential for a healthy life as they digest foods for us and fight hostile bacteria, whereas problems with our gut flora can lead to intestinal diseases, obesity, and even depression.

Like us, all other mammals have been dependent on uncounted species of microbes ever since they roamed the world. But for humans, an additional and very intriguing relationship developed a long time ago: we have been using microbes regularly to our advantage, initially without even knowing it! Ever since we figured out how to make cheeses, yoghurts, bread, pickles, sauerkraut, vinegar, beer, wine, and more, we have unwittingly enrolled bacteria and fungi to assist us. All these foods and drinks would not have been possible without them. But their existence was unknown until the invention of the microscope in the mid-17th century and insightful experimentation by the Dutch scientist and father of microbiology, Antonie van Leuwenhoek.

Two hundred years later, some scientists, including Louis Pasteur, began to realize that some microbes naturally produce compounds we could use for our own benefit. In particular, it was discovered that bacteria and yeast *ferment* carbohydrates into products like lactate, alcohol, or vinegar. For instance, the small bacterium *Lactococcus* that we discussed in Chapter 2 converts sugar into lactate, which prevents yoghurt and cheese from easily spoiling.

A particularly illustrative example is citric acid. As the name indicates, this compound was initially associated with citrus fruit like lemons and oranges, and it turned out to be an excellent preservative for jams, jellies, pickles, and other foods, because its acidity repels many bacteria that could otherwise spoil these foods. Consequently, a flourishing industry at the end of the 19th century produced copious amounts of citric acid from lemons and oranges. Then the American food chemist James Currie observed that a small black fungus by the name of *Aspergillus niger* produced the same citric acid and excreted it into its environment in order to keep other microbes from competing with it for food sources. Currie experimented with different substrates and growth conditions, scaled up production, converted the citric acid into a powder and, by 1917, was able to produce large amounts of citric acid for a fraction of the cost of citrus fruit. In fact, owing to Currie's discovery, the United States became self-sufficient and produced enough citric acid for export to Europe. It is fair to say that Currie's pioneering work optimizing biotechnological production processes make him one of the fathers of the modern field of *metabolic engineering*. Today, over a million tons of citric acid are produced every year for soft drinks and various foods, as well as for use as a starting compound in the chemical, pharmaceutical, and cosmetics industries. Most of this citric acid is produced by carefully selected mutant strains of the *A. niger* mold, which have become valuable and well-protected trade secrets. Different strains are able to grow on a variety of substrates that are cheaper than glucose, including agricultural wastes from processing apples, orange, kiwi, and pineapple peels, cotton waste, and cane molasses.

Since those early days of metabolic engineering, we have learned a lot about biochemistry and molecular biology and now understand at least the basic principles with which microbes convert input compounds into valuable organic chemicals. Two overarching targets have crystallized in the field. One is the large-scale production of *bulk* compounds like industrial ethanol and citric acid, which microbes generate by the millions of liters. The second target is the smaller-scale production of complex, very valuable organic compounds like insulin and amino acids that can be used for prescription drug development or as additives for foods and animal feed. Amino acids pose an interesting challenge. While it is not too difficult to produce most amino acids with methods of analytical chemistry, the result is usually a mixture of two mirror-image forms of the same amino acid, called L and D. This difference may appear to be trivial, but human and animal cells can only use the L-form, whereas the D-form can be toxic, and it is very expensive to separate the two forms cleanly and completely. In contrast to chemical synthesis, bacteria always produce the same form of an amino acid, so that a separation step is not needed. Secondary goals of metabolic engineering are to minimize costs and to maximize yield. The former is often a matter of finding cheap substrates on which the microbes grow with sufficient speed. The latter is complicated, and it is here that systems biology comes in.

In many cases, the chosen bacteria or fungi already produce a compound of interest, but the amount is too small to be of economic value. In other cases, the organism that naturally produces the compound is difficult to grow on a large scale, and the metabolic engineer tries to take its pertinent genes and insert them into a related, more convenient organism, such as the workhorse *Escherichia coli* or the baker's yeast, *Saccharomyces cerevisiae*. Many genetic methods have been developed for such a transfer, but even if the new host accepts and expresses these genes, the yield is almost always rather small. Here, systems biology has the potential to help. Because the desired compound is the output of metabolic pathways, the goal is to alter and reroute parts of the metabolic system toward this compound, while ensuring that the microbe generates enough of its regular metabolites to thrive. It is also necessary either to make the microbes excrete the desired product, for instance by triggering the production of transport proteins, or to design some other strategy of harvesting it.

Three systems approaches have been applied to the task. The first is called *Flux Balance Analysis* (FBA). It attempts to reroute *fluxes* in a targeted manner. At a steady state, none of the metabolite concentrations are changing. Therefore, all fluxes *into* any metabolite pool must exactly balance all fluxes *out of* this pool. Importantly, different combinations of fluxes can satisfy this demand. FBA has been used to reconstruct the metabolic networks of entire organisms and to make predictions of changes in fluxes due to mutations.

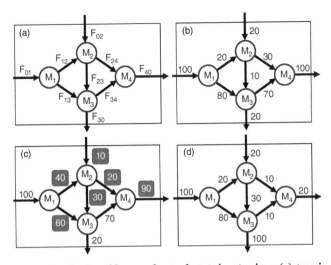

24. The distribution of fluxes at the steady state is not unique: (a) generic illustration of a metabolic network with four metabolites ($M_1, \ldots, M_4$) and nine fluxes ($F_{01}, \ldots, F_{40}$); (b) at the steady state, all fluxes entering a metabolite pool exactly balance those leaving the pool; (c) if flux F_{12} is changed from 20 to 40, other fluxes (grey boxes) can compensate to regain a well-balanced flux-distribution; (d) if the efflux F_{30} is to be maximized under the condition that all fluxes must have a value of at least 10, system-wide rerouting increases its value from 20 to 100.

As an example, consider the small system in Figure 24(a), which contains four metabolite pools and nine fluxes. Total balance at every pool is achieved by the panels (b) and (c), but there are infinitely many others. For an illustration, suppose that M_3 is a valuable compound and that the flux out of M_3 should therefore be maximized. Typically, such optimization problems come with additional constraints. For instance, it might be that the influx into M_1 and M_2 cannot be greater than 20 units and that every internal flux must be at least 10 units in order for the microbe to survive. One solution of this constrained optimization task is given in panel (d), but it is not the only solution. There are many others.

In reality, metabolic pathway systems are much larger, with hundreds of different compounds, and there might be other constraints that must be satisfied. Nonetheless, mathematicians have come up with methods that are very efficient and quickly generate *optimal solutions*, even for very large networks. In these methods, the structure of the flux network is entered into an algorithm, together with constraints and with the *objective*, which here is that the flux should be as strong as possible. With these settings, the algorithm computes the optimal solution, which identifies the highest value of the objective that satisfies all constraints. The method has been applied uncounted times, usually to populations of *E. coli* or other microbes that produce compounds like industrial alcohol, amino acids, or recombinant (genetically manipulated) proteins.

The second method is quite similar to FBA. It is called *Elementary Mode Analysis* and involves sophisticated additional computations and experimental steps. Here are its concepts. Microbes in the rough-and-tumble outside world must be prepared to live under a variety of conditions, digest different substrates, survive through cold weather and droughts, and defend themselves against enemies. All these demands ultimately require metabolic pathways which, for instance, produce lactic acid or citric acid to keep other microbes away. The idea of Elementary Mode Analysis is that we should be able to get rid of many of these pathways, if the microbe did not have to suffer inclement conditions or face enemies. With modest experimental effort, these favorable conditions can be achieved in the laboratory, where food is ample, the temperature is always kept optimal, and everything the microbes need is provided. Thus, the first step is to compute which pathways are necessary for survival under optimal conditions, which pathways or whole metabolic modules are dispensable, and which pathways lead to the desired organic compound the metabolic engineer wants the microbe to produce. Difficult as this sounds, it is indeed possible to identify these pathways with computational methods. Next, all pathways not needed are *knocked out* with methods of modern genetics. The resulting eviscerated strain of the microbe could not possibly survive in the wild for very long, but under optimal laboratory conditions it is actually happy and in turn produces the desired compound, ideally in copious amounts. This method has been spectacularly successful in a number of cases, such as the generation of industrial alcohol from different carbohydrates.

The third method is based on systems of ODEs that account for all metabolites and fluxes, which may change over time, as well as all known regulatory signals. Because the equations tend to be complicated, analysis of these systems is much more difficult. Nonetheless, even though the math is more difficult, we should not ignore regulation, because microbes respond to most artificial changes, such as gene knock-outs or changes in their environments, by trying to compensate for what's missing, and this compensation always involves regulatory signals. A very interesting example illustrating such natural responses is a study in which Japanese researchers disrupted the expression of many genes coding for important enzymes. As a consequence, many mRNAs and proteins displayed quite different concentrations, but the metabolites were almost unchanged. The authors attributed this metabolic stability to regulation that compensated for the perturbations, in addition to the natural, structural robustness of the metabolic system, its redundancy, and the organism's capability of rerouting fluxes in a targeted manner. This example demonstrates how important it is to understand the regulatory and compensatory mechanisms employed by all cells.

The general strategy of the dynamic approach to metabolic engineering is to formulate the entire problem as an ODE model. This model captures reality as well as possible, including compensatory mechanisms, and satisfies all constraints, such as minima and maxima for fluxes and concentrations that the organism can tolerate. Finally, sophisticated computer algorithms are employed to optimize some objective, such as the largest possible production of insulin. In contrast to the earlier methods, this approach takes into account all regulatory signals in the system, as far as they are known.

From a procedural point of view, the algorithm is designed to change some or all enzyme activities within the system and to check whether the objective is improved. Because of the typical number of metabolites and enzymatic processes, millions of combinations are possible, and the algorithm tests so many model settings that, in the end, the best possible profile of enzyme activities is achieved, where the objective is maximized and all constraints are satisfied. Ideally, the metabolic engineer would now be able to manipulate the enzyme activities through a variety of methods so that they match the computed profile and the task of maximizing production is achieved. In reality, many steps in this process are very complicated. First, the ODE model may not include all pertinent components and processes. Second, the computer algorithm may not find the best possible solution, due to technical hurdles. Third, it may not be possible to implement the prescribed solution in the actual microbes. And finally, the microbes can always come up with surprising responses that nobody was able to foresee. So, the problems are manifold, but a general strategy does exist, and it will just be a matter of time before the method bears substantial fruit. A successful demonstration of these methods was the production of the pharmaceutical compound carnitine by *E. coli*.

In principle, all three methods can not only be used with microbes, but can also be applied to plant and mammalian cells, including human cell cultures. However, these eukaryotic cells of 'higher organisms' are incomparably more complicated. For instance, they do not naturally live by themselves in a liquid solution, as bacteria can do, but often need physical contact with other cells and sometimes only grow in single layers, which makes large-scale production much more complicated than with single-cell bacteria or fungi. Nonetheless, there is great interest in developing factories of plant and mammalian cells. For instance, if cultured human cells could be coaxed into synthesizing compounds like insulin, new ways of dealing with diabetes might become feasible.

One promising area of application of mammalian cell culture research is in the production of meat. This *cultured meat* is produced in the laboratory with methods of tissue engineering and famously led to the first lab-grown hamburger in 2013, which was made from strips of muscle that had been grown from cow stem cells and cost thousands of dollars. At this point, the jury is still out on whether cultured meat will eventually taste like real steak or hamburger, whether consumers will accept it, and whether it will become commercially feasible.

Simulating crops

Throughout recorded history, the world's population of humans has been growing rapidly, and experts don't agree on when it will reach a stable number. Meanwhile, all people require food, but the surface of the Earth amenable to agriculture will not increase and may even decrease due to a changing climate and to human activities, and we cannot keep converting wilderness into food-producing agricultural land without possibly causing very severe ecological problems. So, how can we bridge this growing gap? The obvious answer is that we must use the available land more effectively, which in turn means cultivating more productive plants and developing new agricultural practices with better yields. Of course, none of this is new: it has been the core tenet of agriculture for thousands of years. As a case in point, comparing a modern corn cob, which may weigh up to 1½ pounds, with its alleged great-great-grandparent, the Central American 35-gram teosinte, which the indigenous populations ate 9,000 years ago,

25. A stunning success of crop breeding. The top shows the 35-gram fruit of the ancient grass teosinte (*Zea luxurians*), which is alleged to be the precursor of modern corn (maize; *Zea mays*; bottom), where a single cob may weigh 1½ pounds. The middle shows a maize-teosinte hybrid.

leaves no doubt that humankind has made great strides toward producing higher yields (Figure 25). Much of this success was due to selecting the best plants for sowing, and also to a lot of trial and error. While impressive, progress has been slow. Given the current world population growth, we cannot wait for another 9,000 years to produce even bigger corn cobs. The challenge is that creating targeted improvements in crops is difficult for many reasons.

The biggest challenge is that plants are much more complex than microbes. For instance, a spruce tree is believed to have between 50,000 and 60,000 genes. That is about two and a half times the number of genes in the human genome! The number of different metabolites in the plant kingdom has been estimated to be between 200,000 and 1,000,000. Nonetheless, modern methods of genetics can be used for breeding new plants and allow modifications of crops that seemed impossible a decade ago. One complication is that many plants do not have two sets of chromosomes as we do, but often more. For instance, bananas have three sets, potatoes four, wheat has six, and sugarcane eight. In fact, these multiple sets of chromosomes, which are sometimes obtained from other species, are the norm rather than the exception for crops that have been bred throughout the past centuries and millennia. A good example is the genus *Brassica*, which contains cabbages, broccoli, cauliflower, turnips, and other food plants.

Methods of genetics even allow us to follow the history of breeding certain plants. For instance, massive duplications of genes have occurred during the cultivation of rice (*Oryza sativa*) since ancient times, with the result that modern rice varieties contain almost 40,000 genes. However, only 2 to 3 per cent of these genes are unique to the important rice subspecies *Oryza sativa indica* and *Oryza sativa japonica*. The multiple copies of genes make experimentation and modeling cumbersome, because otherwise straightforward, routine techniques like knocking out or knocking down a gene must be done for all these copies. An additional challenge is the fact that the expression of genes in some crop plants is *super-coordinated* in the sense that any changes in gene expression for enzymes in some part of metabolism may also affect enzymes in other parts.

Given this complexity, the hope is that computational methods of ML and systems biology may be able to assist the breeding process. As we discussed before, computers can keep vast quantities of data in memory, and modeling is possible even for very large systems. For instance, network models governing our communication channels span nations and even the globe and reroute phone calls if a pertinent tower somewhere in the world fails. In biology, we have not reached a comparable level of network mastery, but it will only be a matter of time and creativity to develop very large models of the physiology of crops. At this point, some of the same techniques we discussed before have been applied to crops, such as FBA and dynamic modeling with ODEs. Many models have focused on the core feature of plants—photosynthesis—upon which almost every organism depends. Other models have addressed specific metabolic pathways of interest and developed model-driven metabolic engineering strategies. Yet others have been explored to understand the flow of carbon within individual plants.

For agricultural purposes, computational models of single plants may be an academic prerequisite, but it is of much greater interest to be able to simulate whole fields of crops, together with environmental conditions and management decisions. Indeed, some agricultural simulators of this type already exist. A prime example is the Soybean Growth Simulation Model (SoySim), developed by the Institute of Agriculture and Natural Resources at the University of Nebraska at Lincoln. Like any good simulator, SoySim requires inputs, which the user supplies. These inputs are fed into the algorithmic machinery simulating the soy plants under the given input conditions, and SoySim produces output that the researcher or farmer interprets and uses for future decisions.

The inputs to SoySim include data characterizing the planting site and weather conditions, latitude and longitude, as well as the so-called maturity group rating. The latter is an assignment of daily time periods needed by the soy plants to develop mature seeds and corresponds to zones across the country. Additionally required inputs are location-specific and include: evaporation of water from the soil; water loss from plants through transpiration; daily measurements of minimal and maximal temperatures; solar radiation; air humidity; and rainfall. Finally, the user must specify critical aspects of crop management, such as dates of sowing, the first emergence of seedlings, the time the stems stop growing in height, and the plant population density.

The computational machinery behind SoySim accounts for all pertinent processes during the growth of the soybeans and the production of seeds, including normal plant development, the expansion of the total leaf area, accumulation of biomass, and aging. These processes are clustered into functional modules within the plants (Figure 26). The details of these modules need not be discussed here; suffice it to say that each contains subsystems that are all coded into mathematical functions and differential equations.

The output of SoySim consists of dates of the various developmental stages of the plants, the growth dynamics of the plants, seed development, yield potential, as well as water use and recommendations for future management.

Taken together, SoySim simulates the growth of soybeans on a daily basis, from their first emergence to harvest. It assumes that sufficient water and nutrients are available and that there are no losses due to disease or environmental factors. To test the *validity* of SoySim—that is, to assess the reliability of its outputs to different sets of inputs—the simulator was used in Nebraska, Iowa, and Indiana. The results demonstrated reasonable accuracy in terms of above-ground growth and seed yield.

Another simulator of interest is WIMOVAC (*Windows Intuitive Model of Vegetation response to Atmosphere and Climate Change*), which was developed at the Universities of Essex and of Illinois. It is a generic simulator for exploring the responses of vegetation to environmental alterations or changes in light, temperature, carbon dioxide (CO_2), humidity, and other factors and provides an intuitive teaching and analysis tool for crop development. WIMOVAC is solidly based on known biochemical and

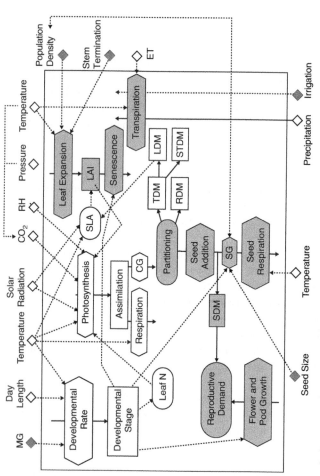

26. Structure of the model on which the soybean growth simulator SoySim is based.

biophysical mechanisms of photosynthesis and focuses specifically on the carbon balance in plants. In particular, it can be used to explore the effects of a wide range of environmental factors over time periods of seconds, days, or even years. Its particular strength is that it spans the range from biochemical knowledge all the way to the level of ecosystems. WIMOVAC accounts for many mechanisms associated with photosynthesis, the uptake of CO_2, light and temperature conditions, and interactions between plants, soil, and atmosphere, and predicts the development of crop canopies and yield in time and space. In fact, for improved accuracy, it computes *microclimate* models for light and temperature profiles in different layers within the canopy. Furthermore, the software is linked to a database of parameter values that permit simulations of different plant species and genotypes without the user having to change the computer code.

The overall structure of WIMOVAC is shown in Figure 27. As in the case of SoySim, many boxes contain modules or subsystems that are encoded in functions and differential equations. As depicted in panel (a), growth depends on environmental factors and the amount of energy needed to reach the next stage of development. The model accounts for different plant organs, which eventually age and contribute carbon and nitrogen for growth. Panel (b) indicates how WIMOVAC calculates the amount of photosynthesis occurring in the canopy in sunlit and shaded leaves. The computations also depend on the anatomical and physiological properties of the leaves.

WIMOVAC may be used to explore various scenarios. An example is the investigation of the effects of changes in temperature, CO_2, or humidity. The simulator shows how plants will adapt to such new conditions and predicts how yields from crops would change. In a test simulation of a poplar forest under current conditions and elevated CO_2, the predicted wood production was consistent with other, independent estimates. The simulation even

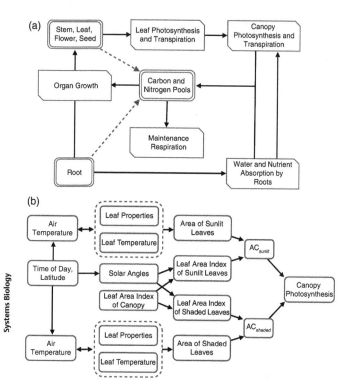

27. Overall structure of WIMOVAC. (a) The growth stage of a plant is determined by the temperature and other environmental factors, as well as the energy needed to reach this stage. Aging of organs leads to recycling of carbon and nitrogen. (b) Computation of photosynthesis in the canopy. Adapted from the literature, where further details can be found.

demonstrated that early canopy closure was more influential than changes in photosynthesis.

While food crops have been studied for hundreds of years, the search for a new assortment of crops for chemical energy production began only a couple of decades ago. This assortment is collectively called *biofuels* and describes plants whose harvested compounds can be converted into fuels for transportation. By far the most used of these compounds at present is ethanol. Gasoline (petrol) in many countries in Europe and the Americas already contains some percentage of it. However, several candidates for alternatives to ethanol are being explored, including the higher-energy alcohol butanol, as well as fatty acids.

Biofuels are interesting from the standpoint of metabolic engineering, because two organisms must be considered: the plant from which the raw material is extracted and the microbes that convert cellulose into ethanol or some other compound. Although this line of applied research is comparatively young, genetic technology and methods of metabolic engineering have made some biofuel production commercially competitive. For instance, factories in Kansas and Iowa started commercial production of ethanol from cellulose in 2014 with an initial annual capacity of 80 million gallons. That sounds like an impressive amount, but constitutes only a minute portion of the ambitious US goal of 1.75 billion gallons per year.

Unfortunately, most of the current production processes use corn or sugarcane, which could otherwise be used to feed humans and animals. Due to this concern, research has shifted toward *second-generation* or *advanced bioenergy* that is to be obtained from inedible plant materials, such as grasses, wood chips, or inedible parts of corn stalks. It is in principle also possible to obtain biofuel materials from algae.

Methods are actually available for extracting ethanol from wood chips or grasses, but the process is so expensive that it is not economically feasible. The root challenge is that cellulose in plants is tightly entangled with a natural polymer called *lignin*, which is the compound that makes wood hard. Except for cellulose itself, lignin is the most abundant biopolymer on Earth, accounting for about 30 per cent of all organic carbon. To make matters even more complicated, lignin is highly resistant to digestion by most microbes, and other organisms cannot really digest it at all. If you think that termites eat wood, it is actually specialized microbial communities in their guts that digest lignin, after it has been chemically altered by the termite.

Thus, one goal is to find bacteria that are really good at digesting wood and converting the cellulose into ethanol. The second goal, for the metabolic engineer on the plant side, is a targeted reduction of the amount of lignin in suitable plants or a change in its chemical composition that makes it easier for the microbes to digest. Of course, one must not entirely eliminate lignin from these plants, because they would simply topple over, but a reduction seems to be a reasonable strategy. In addition to many of the other challenges discussed before, which are due to the complexity of plants, lignin is a peculiar compound, as its natural synthesis contains many steps in which relatively few enzymes convert one intermediate metabolite into another. The consequence is that changing one enzyme with methods of modern genetics automatically changes several reactions, which makes intuitive predictions difficult. Here, systems biology can help by setting up models for lignin production in target plants and exploring computationally how enzymes should be changed to obtain less lignin or a composition of lignin that can be more easily digested by microbes.

In many cases, the creation of simple mathematical models is a good strategy toward understanding the essence of complicated biological phenomena. However, when it comes to being as realistic as possible, simulations are the best—and arguably only—tools for exploring life in its overwhelming complexity. Systems simulation in biology is off to a good start, and the application areas discussed in this chapter demonstrate that simulations are about to reach a level of sophistication that makes them practically useful. Many other areas are sure to follow, and much is yet to be done. The possibilities afforded by simulations are very exciting, and the sky is the limit. Indeed, this aspect of systems biology has the potential to contribute substantially to future solutions of the really big challenges we are facing today and in the future.

Chapter 7
The lawless pursuit of biological systems

If an electrical engineer designs a circuit and it works well, it will work well in any other reasonable circumstances. There is no reason to investigate it further, because the engineer can explain why it works well. Biology is different. If a biologist finds a set of genes that control the response of a bacterium to some environmental stress, it does not mean that the same—or even similar—genes do the same job in a yeast cell. Physicists know that a perpetual motion machine will always be impossible, since it violates the first law of thermodynamics. In stark contrast, biology does not have strict laws governing the limits of life: just recall the extremophiles found living in hot ocean vents (Chapter 6).

The laws of physics are a prerequisite for us to make reliable predictions regarding our surroundings. By extension, if we want to make reliable predictions in biology, we must have laws of biology. The problem is, as we have seen countless times by now, that biological systems are hugely complex and diverse. As a consequence, it is difficult to make true statements covering all organisms on Earth—or even large classes of organisms. This difficulty translates directly into the challenge of identifying rules that govern biological systems. What would such biological rules or laws even look like? In order to develop ideas that might eventually lead to biological laws, let us briefly pause and consider the nature and roles of laws in mathematics and the physical sciences, and perhaps we might see which aspects might be applicable to biological systems.

A mathematical statement is said to be either true or false. And in most cases it is, although not necessarily. The reason for this unexpected and somewhat disturbing caveat is that we first must agree on certain 'ground rules' in the form of *axioms* and the *laws of propositional logic*, before we can tell what is true and what is not. Axioms are statements that we have to accept as true even though they cannot be proved. For example, one axiom states that every natural number (1, 2, 3,…) has a successor. That may sound obvious but it cannot be proven. Another axiom says 'for all natural numbers x, y and z, if $x = y$ and $y = z$, then $x = z$.' An example of a law of propositional logic is the so-called *syllogism*: 'Every person has a natural mother. Emma is a person. Therefore, Emma has a natural mother.' Another law is the *Law of the Excluded Third*, which is crucial in many mathematical proofs. For example: any real number is either a rational number or an irrational number; there is no third option. Modern mathematics uses roughly twenty such laws. Importantly, if we do not accept these axioms and laws as valid, we cannot really discuss math, and a statement that is true for you may be false for me. Mathematicians have pondered these ground rules for a very long time, and essentially every mathematician agrees with them.

The laws of mathematics are abstract constructs of the human mind. In particular, they are independent of data, measurements, or observations. Assuming that we accept the axioms and rules of propositional logic, we can prove or disprove mathematical statements, and that is of great academic and practical value. A minor fly in the ointment is the following: the Austrian mathematician Kurt Gödel showed that there cannot be a contradiction-free 'complete' set of axioms, where every statement can be proven to be true or false. Sometimes we simply cannot decide. But while theoretically important for mathematicians and philosophers, this *incompleteness theorem* does not directly affect our daily lives. For practical purposes, correct math will always be correct math, independent of circumstances, and this general truth is of fundamental importance for our quest to understand the world in which we live.

The laws of physics are distinctly different from the laws of mathematics. Unlike the abstractions and derivations of pure mathematics, physics deals with objects and processes in the real, often less than ideal, world. In physics, and by extension chemistry, laws put boundaries on what is possible, and thereby exclude what cannot happen because it is somehow 'against nature' and therefore never observed. The fresh hot coffee in a mug always gets cooler at room temperature, never hotter; Earth continues to orbit the Sun in a very well defined, predictable ellipse. The laws of physics and chemistry are of immense importance because they allow us to make sweeping statements about the real world. They are true today as they were yesterday, and we expect them to be true tomorrow.

We do not know the origins of the fundamental physical laws under which the Universe operates, and which have enabled the existence of such complex structures as stars, galaxies, and life. Instead, we have been inferring these laws of the physical sciences from observations. A good example is a law stating that, at a constant temperature, the product of the pressure and the volume of an ideal gas is constant. The Anglo-Irish chemist Robert Boyle inferred this law from the results of experiments yielding data points that are redrawn as diamonds in Figure 28. Note that the grey line through the diamonds is not an observation but the result of insight and inference under the assumption that the process is smooth, rather than jumping up and down between the data points. This line, looking quite obvious here, is actually an important ingredient for formulating the law, because the law should be valid for all volumes and pressures, and not just for those that had been

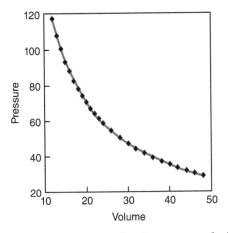

28. Boyle's Law. Observations regarding the pressure and volume of an ideal gas, on which Robert Boyle based his gas law.

measured. Boyle's achievement was the insight that the product of volume and pressure is always constant.

Simple as it is, the example demonstrates the value of a model, law, and theory. Without these three, Boyle's observation would consist of a list of many statements, one for each data point, of the type 'for this gas and under the given conditions: if the volume is 20, then the pressure is about 70'; 'for this gas and under the given conditions: if the volume is 30, then the pressure is about 47;…' In contrast to this genuinely incomplete list, the conceptual model relating volume and pressure compresses all these statements into a single law: 'for this gas and under the given conditions, the product of volume and pressure is constant at ~1400.' A comprehensive theory of gases would further compress this body of information by explaining how the number 1,400 relates to the conditions of the ideal gas and, possibly, how the relationships would change for real, non-ideal gases. Knowing this theory, all future volume–pressure combinations for gases would be exactly predictable.

Similarly, Newton's law of gravity is not important because of apples that he allegedly saw falling from trees, but because it applies (almost) universally, even for the sun and moon and stars. 'Almost,' because, just as other laws in physics, Boyle's and Newton's laws are true for certain domains of conditions, but not outside these domains. In Boyle's case, the product is anything but constant for liquids, which are in most cases incompressible, while in Newton's case, the law of gravity becomes slightly inaccurate when applied at the cosmological scale, because it is really a special case of Einstein's general relativity theory, which has a wider explanatory scope.

Notwithstanding this caveat, we are able to make predictions regarding physical processes and even regarding very large systems with enormous numbers of such processes, because we understand the laws of physics, as well as their limitations. Chapter 6 mentioned the Boeing 787 Dreamliner, an airplane with 2.3 million parts that support countless, often complicated processes. In spite of this complexity, we can make such accurate predictions regarding the functioning of the plane that we entrust it with thousands of lives every day. All these predictions are ultimately based on the laws of physics, which are collectively the foundation of engineering. This certainty is also the basis for explaining why and how something happens in the world of physics and chemistry. Indeed, it is this certainty that something is always correct, and the reliable predictability resulting from this certainty, that makes the laws of physics and chemistry so enormously valuable. The laws of physics and chemistry give us the confidence to ride bikes and mass-produce reliable materials. They have also made it possible to place rovers on the surface of Mars that move around and send us astonishing photographs.

Returning now to biology, we must ask: are there laws that give us the same degree of confidence? Clearly, all laws of physics must hold for biological systems as well, because these systems are parts of the physical world. That argument is certainly valid, but it does not help us much, because even relatively simple biological phenomena involve so many physical processes simultaneously that we cannot tease them apart and instead become overwhelmed. As just one example, the division of a bacterium into two daughter cells involves uncounted molecular components and processes that operate in time and space, and we often cannot conclusively explain which components are to blame if the division runs into problems.

Biology does have a few laws that permit us to gauge their value and impact, and explain the strong desire to formulate more, but these laws are rare. One law was proposed by the 19th century German physician Rudolf Virchow, who stated that 'every cell comes from a cell.' Except for the first cell ever, we will probably agree with that, as long as synthetic biology cannot generate cells from their building blocks. However, while true, one will have to admit that this law does not have a lot of practical punch, for instance in comparison with Maxwell's equations of electromagnetism.

Another high-level law is the *competitive exclusion principle* that is attributed to the 20th-century Russian biologist Georgii Gause, although it is similar to ideas proposed by Charles Darwin several decades before. Gause's law states that two species competing for exactly the same limited resource cannot truly coexist: any advantage one of the two species has will lead to dominance of this species and to the decline, extinction, or evolutionary change of the other species.

A much more potent set of rules with far-reaching consequences is the *genetic code* or the *codon law of genetics*, which has been called the *Rosetta Stone of Life*. It is a matter of semantics whether the genetic code should be called a law or not; it is very important either way. The code pertains to nucleotides, the building blocks of DNA and RNA, and states that each combination of three neighboring nucleotides, called a *codon*, always codes for one particular amino acid in the protein that is translated from the RNA, independent of the organism or species. For instance, the triplet AAA within the DNA molecule of a gene always signals that, if the gene is transcribed and translated into a protein, this protein will contain the amino acid lysine in the corresponding position. Because this law is believed to be universal, we know immediately what every DNA codon within any gene of any species signifies. Note that the order of nucleotides is important, so that the combination GAC always codes for leucine, while AGC codes for serine. A direct consequence of the codon law is information regarding proteins: because each codon universally represents a specific amino acid, knowledge of an organism's gene sequences immediately implies that we know with confidence which proteins this organism is able to produce.

Without the genetic code, we would need to elucidate anew the rules by which each species (or maybe even every organism) would store information for creating proteins from its DNA. So, this law is of fundamental importance because of its universal predictability. Are there any limits, like with the Newtonian law of gravity? Yes, there are, but very few. For instance, the codon TAC has a dual function: it may code for the amino acid methionine, but it may also trigger the start of RNA translation into protein. We do not fully understand the control mechanisms determining the fate of TAC in a given situation. Similarly, the codon ACT may code for cysteine as well as for the rare amino acid selenocysteine, and it may furthermore signal the termination of translation. Finally, there are some altered codon assignments in a small number of rare species. So, yes, there are a few exceptions, but the genetic code is true in the vast majority of cases, and that is of extreme value for genetic research and has unlimited consequences for medicine, agriculture, drug development, and a host of other grand challenges.

The codon law is a shining exception within an otherwise mostly lawless world of biology. In physics, we can answer many 'why' questions due to its laws. For instance, when we ask 'why did the apple fall off the tree?' the answer is 'because of gravity,' and every educated person will agree with us. But if we ask 'why is a Monarch butterfly orange and black?' we cannot come up with a satisfactory answer that is indisputable.

One reason for the scarcity of unifying laws in biology is again the enormous complexity and variability of organisms. Reminiscent of the famous quote of the Greek philosopher Heraclitus of Ephesus, that it is impossible to step into the same river twice, every organism is at least slightly different from any other organism so that many commonalities are overshadowed by a host of exceptions. In other words, every organism functions in a slightly different manner, and as a consequence, any laws—if they exist—are either too complicated for us to infer from data, like Boyle did, or they are masked by a myriad of other complicated processes that are superimposed on the process we are trying to understand.

Further confounding the search for potent laws, and indeed puzzling, is the *emergence of system properties* that cannot be explained based on any of the system components alone but only through their interactions. This type of emergence is a central topic of systems biology, but it is difficult to explain. As a rather simple example, we considered in Chapter 2 a system with only two components and three regulatory signals (Figure 3), which is able to generate different types of responses, including oscillations. What is causing these oscillations, if neither component can oscillate without the other? There are uncounted such emergent phenomena in biological systems, many overwhelmingly complicated, such as cognition, thought, and memory, which result from systems of simple nerve cells and their connections. A very challenging, intrinsic problem with the emergence of such new properties is the fact that causality cannot be identified, and without causality, it seems difficult to formulate a law.

Emergence in the extreme can be found in seemingly simple systems that display unpredictable *chaotic behaviors*, which can only be assessed by solving the equations defining the system. Which parts of the system actually cause this chaos is impossible to say. For example, let's look at a system of three equations:

$$\frac{dX}{dt} = sin(Y)$$
$$\frac{dY}{dt} = sin(Z)$$
$$\frac{dZ}{dt} = sin(X) \tag{7}$$

This system does not contain any 'weird' components, but only simple sine functions. Solving the differential equations with start values $X(0) = 0$, $Y(0) = 4$, $Z(0) = 0.5$ at time $t = 0$ leads to trajectories that are typical for this system but impossible to foresee (Figure 29(a)). Maybe even more puzzling, if we change the start value of any variable ever so slightly, for instance, by setting $Z(0) = 0.51$, the trajectories are totally different (Figure 29(b)). Such fickleness certainly does not help the search for clear underlying principles or laws. One could imagine that the system needs some time to settle into some pattern, but that is not the case: if we solve the system in Figure 29(b) to time point $t = 500,000$, it behaves as erratically throughout as it did at the beginning (Figure 29(c)). The weirdness of the chaotic oscillator becomes even more evident when we plot Y against X and Z (Figure 29(d)).

One could surmise that the choice of sine functions was needed to create the chaotic behavior, but that is not so: Even differential equations typically used for describing the interactions among different species living together in the same ecological space can generate unpredictable, chaotic trajectories. One might also think that biology would not allow chaotic systems, but that is far from

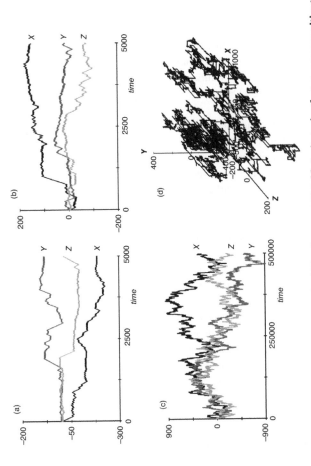

29. **Chaos!** The seemingly simple equations in Eq. (7) lead to unpredictable trajectories that are very sensitive to changes in start values: (a) trajectories from start values $X(0) = 0$, $Y(0) = 0$, $Z(0) = 0.5$; (b) trajectories from start values $X(0) = 0$, $Y(0) = 4$, $Z(0) = 0.51$; (c) even over longer time periods, the system response does not settle down; (d) a plot of one variable against the other two (from panel (c)) shows the weirdness of the system even more clearly.

the truth, if we look closely. In fact, some scientists claim that the loss of chaotic processes in our bodies is a sign of frailty and disease.

So, how can we move forward? Several strategies are currently being pursued, but none of them is a silver bullet. At the one extreme, laws can be formulated, based on intuition, at a very high level, for instance by stating that 'every organism or cell operates toward survival, either of itself or of the population to which it belongs.' The theory of evolution is a conceptual framework containing high-level laws.

At the other extreme, theories have been developed to focus on systems that are close to abstract mathematical structures, such as graphs, for which much theory has been developed. An outcome from this approach is the observation that many biological networks display the *small-world property*. This pattern of network connectivity, recall, resembles in structure the global network of airports and flights between them: the arrangement consists of relatively few major *hubs* and many more secondary airports and allows passengers to reach any airport from any other airport with far fewer connecting flights than a randomly connected network would require. Metabolism appears to have a similar organization: a few *hub* metabolites, such as water and ATP, are involved in many biochemical reactions, whereas most other metabolites are involved in only a few. It is debatable whether this small-world property amounts to a law, as there are always exceptions. However, it is possible that it could amount to a law in some limited biological domains.

Also focusing on network structures is the quite complicated framework of *Chemical Reaction Network Theory*. This line of investigation attempts to frame chemical reaction systems in such a manner that some of their features can be predicted and proven with mathematical rigor. As a specific example, the theory attempts to identify criteria in chemical reaction networks that permit the prediction of how many steady states are possible in these networks. The search for laws in this case resembles the approach of mathematics.

If we want to follow the approach of physics instead, how can we infer new biological laws or at least candidates for laws from observations? Because biology is so noisy, we might actually start by aiming a bit lower than physics and concede from the beginning that such laws would not have to be absolutely universal or that they might have to be restricted to some relatively narrow domains. Addressing the former, we could scout for 'fuzzy' laws of the type: we cannot be absolutely sure, but X will happen in this situation with ~95 per cent probability. Even fuzzy laws would clearly be quite beneficial.

Let us explore the role of systems biology in this context by revisiting the case of a linear metabolic pathway with feedback inhibition (Figure 30), which we studied before in Chapters 2 and 5. The biological purpose of this very frequently observed set-up is that a reduced amount of material should flow into the pathway if enough end product is already available. An obvious question is: why is there not inhibition from B_1, or sequential inhibition back from B_4 (Figure 31), or some entirely different design?

This type of question is difficult to assess with laboratory experiments, because it is difficult to make a metabolite send a feedback signal to some process in the system, and even if the task

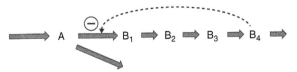

30. Feedback inhibited pathway that branches at metabolite A and then is linear. The more B_4 is available, the stronger is the feedback signal. The result is reduced flow toward B_1 and more material branching off at A.

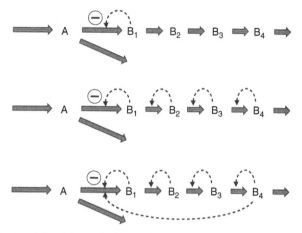

31. Search for design principles. Alternative modes of restricting influx into a pathway, if enough material is available already. Many other options are possible.

could be accomplished, the question would remain as to what else in the organism might have changed. Moreover, there are many combinations of possible signals, especially if positive signals are allowed as well. By contrast, computational systems biology has effective tools to design and analyze mathematical models that accurately capture the functionality of the original pathway (Figure 30) as well as all possible alternatives, including those in Figure 31. Thus, instead of trying to answer 'why' questions experimentally, we can do it rigorously and efficiently with mathematical models.

In the first step, the models are set up in such a way that they are as similar as possible with respect to functions and parameter values and merely differ in the existence or absence of feedback signals, whether they are inhibiting or activating. In the next step, a list of advantageous and disadvantageous features of such a pathway is composed. For instance, it will be considered an advantage if the system responds quickly and appropriately to new situations. By contrast, there is no need to accumulate intermediates like B_2 and B_3, because they would fill the cell without a specific purpose outside being precursors of B_4. So, their accumulation would be considered a drawback. Once the list is assembled, all model variants are tested against this list, and the best-performing model is identified. For the given example, this was actually done, and the victor is the model in Figure 30.

Systems biologists who pursue this line of thinking are interested in *design* and *operating principles*, where the fundamental question is: why do we observe a particular design (or procedure) X rather than an alternative design (or procedure) Y, which seems just as plausible? The goal is to prove with mathematical rigor why, or under what circumstances, a particular design or procedure is optimal, or at least better than the alternatives. The answers to these questions are becoming increasingly important as synthetic biology is getting better at modifying parts of cells or even creating pathway systems within cells. If we create new systems, we should remember that nature has had 3.5 billion years to try out different options and identify the best possible designs among numerous alternatives.

Ultimately these types of analyses will associate more and more design and operating features with frequent tasks that a cell or organism has to accomplish, and the hope is that these associations will turn into laws, which at first may only cover small niches, but increasingly become more general. Considering physics and the enormous successes of its applications, the sky would be the limit if biology were to become reliably predictable. Just imagine how that would affect medical interventions and all other applications of biology.

As systems biology matures as a field of investigation, some of the freedoms of its infancy will fade and catalogs of rules will be established for how to approach a given task. Eventually, this standardization will lead to streamlined and partially automated research procedures that work as pipelines from raw data to curated data, to patterns and associations revealed by ML, and ultimately to explanatory and predictive dynamic models. Enough experience with these pipelines will suggest commonalities within and among biological domains, and these will eventually morph into laws. The same trend of development occurred in physics over the centuries, with spectacular results. There is every reason to be confident that biological systems research will follow a similar path, if bright scientists set their minds to it. The future of systems biology is challenging for sure. It is also enormously promising and, without doubt, exciting all the way.

References

Chapter 1: What is systems biology all about?

Savageau, M. A., Reconstructionist molecular biology. *New Biol.* **3**, 190–7, 1991.

Savageau, M. A., The challenge of reconstruction. *New Biol.* **3**, 101–2, 1991.

Voit, E. O., Dimensions of the scientific method. *PLoS Comp. Biol.* **15(9)**, e1007279, 2019.

Chapter 2: Exciting new puzzles

Dolatshahi, S., L. L. Fonseca, and E. O. Voit, New insights into the complex regulation of the glycolytic pathway in *Lactococcus lactis*. *Molecular Biosystems* **12(1)**, 23–47, 2016.

Fonseca, L. L., C. Sánchez, H. Santos, and E. O. Voit, Complex coordination of multi-scale cellular responses to environmental stress. *Mol. BioSyst.* **7(3)**, 731–41, 2011.

Holland, J. H., *Complexity: A Very Short Introduction*, Oxford University Press, 2014.

Rintamäki, H., Human responses to cold. *Alaska Medicine* **49(2 Suppl.)**, 29–31, 2007.

Saadatpour, A., I. Albert, and R. Albert, Attractor analysis of asynchronous Boolean models of signal transduction networks. *J. Theor. Biol.* **266**, 641–56, 2010.

Voit, E. O., A. R. Neves, and H. Santos. The intricate side of systems biology. *Proc. Nat. Acad. Sci.*, **103(25)**, 9452–7, 2006.

Voit, E. O., Modeling metabolic networks using power-laws and S-systems. *Essays in Biochemistry* **45**, 29–40, 2008.

Chapter 3: The —omics revolution

Carnap, R. *Philosophical Foundations of Physics*. Basic Books, Inc., 1966, ch. 11.

Horgan, R. P., and L. C. Kenny, 'Omic' technologies: Genomics, transcriptomics, proteomics and metabolomics. *The Obstetrician & Gynaecologist* **13**: 189–95, 2011.

Tukey, J. W., *Exploratory Data Analysis*. Addison-Wesley, 1977.

Voit, E. O., Dimensions of the scientific method. *PLoS Comp. Biol.* **15**(**9**), e1007279, 2019.

Chapter 4: Computational systems biology

Caldarelli, G., and M. Catanzaro, *Networks: A Very Short Introduction*, Oxford University Press, 2012.

Theobald, O., *Machine Learning for Absolute Beginners: A Plain English Introduction* (Second Edition), Scatterplot Press, 2017.

Voit, E. O., The best models of metabolism. *WIREs Syst. Biol. Med.* e1391, 2017.

Voit, E. O., *A First Course in Systems Biology*, Garland Press, 2017.

Voit, E. O., Dimensions of the scientific method. *PLoS Comp. Biol.* **15**(**9**), e1007279, 2019.

Chapter 5: Interdependencies of biological systems

Fonseca, L. L., P. W. Chen, and E. O. Voit, Canonical modeling of the multi-scale regulation of the heat stress response in yeast. *Metabolites* **2**(**1**), 221–41, 2012.

KEGG (Kyoto Encyclopedia of Genes and Genomes). https://www.genome.jp/kegg/

Kung, S. H., S. Lund, A. Murarka, D. McPhee, and C. J. Paddon, Approaches and recent developments for the commercial production of semi-synthetic artemisinin. *Front. Plant Sci.* **9**, 87, 2018.

Liberti, M. V., and J. W. Locasale, The Warburg effect: How does it benefit cancer cells? *Trends Biochem Sci.* **41**(**3**): 211–18, 2016.

Øyås, O., and J. Stelling, Genome-scale metabolic networks in time and space. *Curr. Opin. Syst. Biol.* **8**, 51–8, 2018.

Singh, D., and A. Laxmi, Transcriptional regulation of drought response: A tortuous network of transcriptional factors. *Front. Plant Sci.* **6**, 895, 2015.

Vera, J., and O. Wolkenhauer, A system biology approach to understand functional activity of cell communication systems. *Methods. Cell Biol.*, **90**, 399–415, 2018.

Voit, E. O., *A First Course in Systems Biology*, Garland Press, 2017.

Chapter 6: Simulators

Anaya-Reza, O., and T. Lopez-Arenas, Comprehensive assessment of the L-lysine production process from fermentation of sugarcane molasses, *Bioprocess Biosyst. Eng.* **40**(7), 1033–48, 2017.

Davis, J. D., C. M. Kumbale, Q. Zhang, and E. O. Voit, Dynamical systems approaches to personalized medicine. *Curr. Opin. Biotechnol.* **58**, 168–74, 2019.

Faraji, M., and E. O. Voit, Improving bioenergy crops through dynamic metabolic modeling. *Processes* **5**, 61, 2017.

Foubister, V., Genes predict childhood leukemia outcome. *Drug Discovery Today* **10**, 812, 2005.

Geib, S. M., T. R. Filley, P. G. Hatcher, K. Hoover, J. E. Carlson, M. del Mar Jimenez-Gasco, A. Nakagawa-Izumi, R. L. Sleighter, and M. Tien, Lignin degradation in wood-feeding insects. *Proc. Nat. Acad. Sci. U.S.A.* 105(35), 12932–7, 2008.

Ishii, N., K. Nakahigash, T. Baba, M. Robert, T. Soga, A. Kanai, T. Hirasawa, M. Naba, K. Hirai, A. Hoque, P. Y. Ho, Y. Kakazu, K. Sugawara, S. Igarashi, S. Harada, T. Masuda, N. Sugiyama, T. Togashi, M. Hasegawa, Y. Takai, K. Yugi, K. Arakawa, N. Iwata, Y. Toya, Y. Nakayama, T. Nishioka, K. Shimizu, H. Mori, and M. Tomita, Multiple high-throughput analyses monitor the response of *E. coli* to perturbations. *Science* **316** (**5824**), 593–7, 2007.

Kulkarni, R., Metabolic engineering: Biological art of producing useful chemicals. *Resonance* **21**, 233–7, 2016.

Lee, K. Y., J. M. Park, T. Y. Kim, H. Yun, and S. Y. Lee, The genome-scale metabolic network analysis of *Zymomonas mobilis* ZM4 explains physiological features and suggests ethanol and succinic acid production strategies. *Microb. Cell Factories* **9**, 94, 2010.

Marshall-Colon, A., S. P. Long, D. K. Allen, G. Allen, D. A. Beard, B. Benes, S. von Caemmerer, A. J. Christensen, D. J. Cox, J. C. Hart, P. M. Hirst, K. Kannan, D. S. Katz, J. P. Lynch, A. J. Millar, B. Panneerselvam, N. D. Price, P. Prusinkiewicz, D. Raila, R. G. Shekar, S. Shrivastava, D. Shukla, V. Srinivasan, M. Stitt, M. J. Turk, E. O. Voit, Y. Wang, X. Yin, and X.-G. Zhu,

Crops in silico: Generating virtual crops using an integrative and multi-scale modeling platform. *Frontiers in Plant Science* **8**, 2017.

Meadows, A. L., R. Karnik, H. Lam, S. Forestell, and B. Snedecor, Application of dynamic flux balance analysis to an industrial *Escherichia coli* fermentation. *Metabol. Eng.* **12**, 150–60, 2010.

Orth, J. D., I. Thiele, and B. Ø. Palsson, What is flux balance analysis? *Nat. Biotechnol.* **28**, 245–8, 2010.

Øyås, O., and J. Stelling, Genome-scale metabolic networks in time and space. *Curr. Opin. Syst. Biol.* **8**, 51–8, 2018.

Santos, G., J. A. Hormiga, P. Arense, M. Canovas, and N. V. Torres, Modelling and analysis of central metabolism operating regulatory interactions in salt stress conditions in a L-carnitine overproducing *E. coli* strain. *PLoS One* **7**(4), e34533, 2012.

Song, Q., D. Chen, S. P. Long, and X. G. Zhu, A user-friendly means to scale from the biochemistry of photosynthesis to whole crop canopies and production in time and space—Development of Java WIMOVAC. *Plant, Cell & Environment* **40**, 51–5, 2017.

Stephanopoulos, G., A. A. Aristidou, and J. Nielsen, *Metabolic Engineering: Principles and Methodologies*, Academic Press, 1998.

Trinh, C. T., A. Wlashin, and F. Srienc, Elementary mode analysis: A useful metabolic pathway analysis tool for characterizing cellular metabolism. *Appl. Microbiol. Biotechnol.* **81**, 813–26, 2009.

Torres, N. T., and E. O. Voit, *Pathway Analysis and Optimization in Metabolic Engineering*, Cambridge University Press, 2002.

Voit, E. O., Z. Qi, and S. Kikuchi, Mesoscopic models of neurotransmission as intermediates between disease simulators and tools for discovering design principles. *Pharmacopsychiatry* **45**(**Suppl. 1**), S22–30, 2012.

Voit, E. O., *A First Course in Systems Biology*, Garland Science, 2017.

Wittig, V. E., C. J. Bernacchi, X. G. Zhu, C. Calfapietra, R. Ceulemans, P. Deangelis, B. Gielen, F. Milglietta, P. B. Morgan, and S. P. Long, Gross primary production is stimulated for three *Populus* species grown under free-air CO_2 enrichment from planting through canopy closure. *Global Change Biology* **11**, 644–56, 2005.

Chapter 7: The lawless pursuit of biological systems

Alves, R., and M. A. Savageau, Effect of overall feedback inhibition in unbranched biosynthetic pathways. *Biophys. J.* **79**, 2290–304, 2000.

Arceo, C. P., E. C. Jose, A. Marin-Sanguino, and E. R. Mendoza, Chemical reaction network approaches to Biochemical Systems Theory. *Mathem. Biosc.* **269**, 135–52, 2015.

Goldberger, A. L., Is the normal heartbeat chaotic or homeostatic? *Physiology* **6**(2), 87–91, 1990.

Lipsitz, L. A., Physiological complexity, aging, and the path to frailty. *Sci. Aging Knowledge Environm.* pe16, 2004.

Lobanov, A. V., A. A. Turanov, D. L. Hatfield, and V. N. Gladyshev, Dual functions of codons in the genetic code. *Crit. Rev. Biochem. Mol. Biol.* **45**, 257–65, 2010.

O'Connor, T., and H. Y. Wong, Emergent properties, in E. N. Zalta (ed.), *The Stanford Encyclopedia of Philosophy* (Summer 2015 Edition), https://plato.stanford.edu/archives/sum2015/entries/properties-emergent (accessed 2019).

Sprott, J. C., J. A. Vano, J. C. Wildenberg, M. B. Anderson, and J. K. Noel, Coexistence and chaos in complex ecologies. *Physics Letters A* **335**, 207–12, 2005.

Thomas, R., Deterministic chaos seen in terms of feedback circuits: Analysis, synthesis, 'labyrinth chaos.' *Int. J. Bifurcation Chaos* **9**, 1889–905, 1999.

Voit, E. O., Design principles and operating principles: The yin and yang of optimal functioning. *Math. Biosci.* **182**, 81–92, 2003.

Further reading

Conceptual books

Eberhard O. Voit, *The Inner Workings of Life. Vignettes in Systems Biology*, Cambridge University Press, 2016.

John H. Holland, *Complexity: A Very Short Introduction*, Oxford University Press, 2014.

Albert-László Barabási, *Linked: How Everything is Connected to Everything Else and What it Means for Business, Science, and Everyday Life*, Penguin Books Ltd, 2013.

David S. Goodsell, *The Machinery of Life*, Springer Science+Business Media, 2010.

Denis Noble, *The Music of Life: Biology Beyond Genes*, Oxford University Press, 2006.

Terence Allen and Graham Cowling, *The Cell: A Very Short Introduction*, Oxford University Press, 2011.

Introductory technical books

Eberhard O. Voit, *A First Course in Systems Biology*, Garland Press, 2017.

Edda Klipp, Wolfram Liebermeister, Christoph Wierling, and Axel Kowald, *Systems Biology: A Textbook*, Wiley VCH, 2016.

A. J. Marian Walhout, Marc Vidal, and Job Dekker (eds), *Handbook of Systems Biology: Concepts and Insights*, Academic Press, 2013.

Index

For the benefit of digital users, indexed terms that span two pages (e.g., 52–53) may, on occasion, appear on only one of those pages.

Conscience
A Very Short
Introduction
Paul Strohm

In the West conscience has been relied upon for two
thousand years as a judgement that distinguishes right from
wrong. It has effortlessly moved through every period division
and timeline between the ancient, medieval, and modern. The
Romans identified it, the early Christians appropriated it, and
Reformation Protestants and loyal Catholics relied upon its
advice and admonition. Today it is embraced with equal
conviction by non-religious and religious alike. Considering its
deep historical roots and exploring what it has meant to
successive generations, Paul Strohm highlights why this
particularly European concept deserves its reputation as 'one
of the prouder Western contributions to human rights and
human dignity throughout the world.

www.oup.com/vsi

AFRICAN HISTORY
A Very Short Introduction
John Parker & Richard Rathbone

Essential reading for anyone interested in the African continent
and the diversity of human history, this *Very Short Introduction*
looks at Africa's past and reflects on the changing ways it has
been imagined and represented. Key themes in current thinking
about Africa's history are illustrated with a range of fascinating
historical examples, drawn from over 5 millennia across this
vast continent.

'A very well informed and sharply stated historiography...should
be in every historiography student's kitbag. A tour de force...it
made me think a great deal.'

Terence Ranger,
The Bulletin of the School of Oriental and African Studies

CANCER
A Very Short Introduction
Nick James

Cancer research is a major economic activity. There are constant improvements in treatment techniques that result in better cure rates and increased quality and quantity of life for those with the disease, yet stories of breakthroughs in a cure for cancer are often in the media. In this *Very Short Introduction* Nick James, founder of the CancerHelp UK website, examines the trends in diagnosis and treatment of the disease, as well as its economic consequences. Asking what cancer is and what causes it, he considers issues surrounding expensive drug development, what can be done to reduce the risk of developing cancer, and the use of complementary and alternative therapies.

GEOGRAPHY
A Very Short Introduction
John A. Matthews & David T. Herbert

Modern Geography has come a long way from its historical
roots in exploring foreign lands, and simply mapping and naming
the regions of the world. Spanning both physical and human
Geography, the discipline today is unique as a subject which
can bridge the divide between the sciences and the
humanities, and between the environment and our society.
Using wide-ranging examples from global warming and oil,
to urbanization and ethnicity, this *Very Short Introduction* paints
a broad picture of the current state of Geography, its subject
matter, concepts and methods, and its strengths and
controversies. The book's conclusion is no less than
a manifesto for Geography' future.

> 'Matthews and Herbert's book is written- as befits the VSI series- in
> an accessible prose style and is peppered with attractive and
> understandable images, graphs and tables.'
>
> Geographical.

DESERTS
A Very Short Introduction
Nick Middleton

Deserts make up a third of the planet's land surface, but if you picture a desert, what comes to mind? A wasteland? A drought? A place devoid of all life forms? Deserts are remarkable places. Typified by drought and extremes of temperature, they can be harsh and hostile; but many deserts are also spectacularly beautiful, and on occasion teem with life. Nick Middleton explores how each desert is unique: through fantastic life forms, extraordinary scenery, and ingenious human adaptations. He demonstrates a desert's immense natural beauty, its rich biodiversity, and uncovers a long history of successful human occupation. This *Very Short Introduction* tells you everything you ever wanted to know about these extraordinary places and captures their importance in the working of our planet.

SLEEP
A Very Short Introduction
Russell G. Foster & Steven W. Lockley

Why do we need sleep? What happens when we don't get
enough? From the biology and psychology of sleep and the
history of sleep in science, art, and literature; to the impact of
a 24/7 society and the role of society in causing sleep disruption,
this *Very Short Introduction* addresses the biological and
psychological aspects of sleep, providing a basic understanding
of what sleep is and how it is measured, looking at sleep
through the human lifespan and the causes and consequences
of major sleep disorders. Russell G. Foster and Steven
W. Lockley go on to consider the impact of modern society,
examining the relationship between sleep and work hours,
and the impact of our modern lifestyle.

www.oup.com/vsi

FORENSIC SCIENCE
A Very Short Introduction
Jim Fraser

In this Very Short Introduction, Jim Fraser introduces the concept of forensic science and explains how it is used in the investigation of crime. He begins at the crime scene itself, explaining the principles and processes of crime scene management. He explores how forensic scientists work; from the reconstruction of events to laboratory examinations. He considers the techniques they use, such as fingerprinting, and goes on to highlight the immense impact DNA profiling has had. Providing examples from forensic science cases in the UK, US, and other countries, he considers the techniques and challenges faced around the world.

An admirable alternative to the 'CSI' science fiction juggernaut...Fascinating.

William Darragh, Fortean Times

www.oup.com/vsi

SOCIAL MEDIA
Very Short Introduction

Join our community
www.oup.com/vsi

- Join us online at the official Very Short Introductions **Facebook** page.
- Access the thoughts and musings of our authors with our online **blog**.
- Sign up for our monthly **e-newsletter** to receive information on all new titles publishing that month.
- Browse the full range of Very Short Introductions online.
- Read **extracts** from the Introductions for free.
- Visit our library of **Reading Guides**. These guides, written by our expert authors will help you to question again, why you think what you think.
- If you are a teacher or lecturer you can order inspection copies quickly and simply via our website.

ONLINE CATALOGUE
A Very Short Introduction

Our online catalogue is designed to make it easy to find your ideal Very Short Introduction. View the entire collection by subject area, watch author videos, read sample chapters, and download reading guides.